FREE MATH TIPS GUIDE TO BOOST YOUR AFOQT SCORE

Scoring well on the two math sections of the AFOQT will make you a more competitive applicant.

To help maximize your score on the test, we've put together a free math tips guide for you.

In this book are two chapters dedicated to the math sections, but this additional math tips guide will further increase your math test preparation skills. Test prep and test taking are skills that can be optimized, and this math tips guide will sharpen your study skills on these sections.

To get your free math tips guide, email math@militaryprepacademy.com with "Math Tips" in the subject line and include the following information in the body of the email:

- The title of your study guide
- Your product rating on a scale of 1 to 5, with 5 being the highest rating
- Your feedback about your study guide. What do you like or dislike about it?
- Your full name and shipping address to receive your math tips guide

Your feedback helps us serve you better.

If you have any questions, send us an email at math@militaryprepacademy.com.

We're here to support you!

AFOQT Study Guide 2022-2023

AIR FORCE OFFICER QUALIFYING

TEST PREP BOOK

Military Prep Academy

Copyright © 2019 by Military Prep Academy

All rights reserved. No part of this publication may be reproduced, stored in a retrieval system, or transmitted in any form or by any means whatsoever without the prior written permission of the publisher, except for brief quotations in critical reviews or articles.

Printed in the United States of America

Paperback
ISBN: 9798559395401

TABLE OF CONTENTS

Introduction 7

Verbal Analogies 9
   Practice Test 17
   Answer Guide 22

Arithmetic Reasoning 25
   Practice Test 64
   Answer Guide 71

Word Knowledge 75
   Practice Test 84
   Answer Guide 89

Math Knowledge 91
   Practice Test 120
   Answer Guide 128

Reading Comprehension 139
   Practice Test 148
   Answer Guide 158

Situational Judgment 161
   Practice Test 162
   Answer Guide 175

Self-Description Inventory 179

Physical Science 181
   Practice Test 230
   Answer Guide 235

Table Reading 239
   Practice Test 242
   Answer Guide 256

Instrument Comprehension **261**
    Practice Test 264
    Answer Guide 277

Block Counting **279**
    Practice Test 282
    Answer Guide 292

Aviation Information **295**
    Practice Test 311
    Answer Guide 315

Exam Success 2.0 . . . . **317**
    10 Tips for Successful Test-Taking Preparation. 317
    10 Tips for Successful Test-Taking. 323

INTRODUCTION

The Air Force Officer Qualifying Test is a standardized test that the Air Force uses to assess your quantitative and verbal aptitude relevant to potential career fields. The test is used to qualify for careers such as pilot, combat systems officer, and air battle manager training. Also, the test is used as a determining factor for program admittance into Officer Training School and Air Force Reserve Officer Training Corps.

Each chapter in this study guide is dedicated to a subtest of the AFOQT. There are 12 subtests in the AFOQT, hence there are 12 chapters in this study guide. Each chapter covers the potential content in each section of the AFOQT. At the end of each chapter, there is a practice test with a corresponding answer guide. It is recommended that you take each practice test with a timer and abstain from looking at the answer guide until your time runs out. After the time limit has been reached, you may compare your answers with the answer key. This will give you an awareness of your current level of competency and give insight into the areas that require more focus.

Below you will see how many questions and minutes you will be given on each subtest of the AFOQT:

Subtest	Test Time (minutes)	Items
Verbal Analogies	8	25
Arithmetic Reasoning	29	25
Word Knowledge	5	25
Math Knowledge	22	25
Reading Comprehension	38	25
Situational Judgment	35	50
Self-Description Inventory	45	240
Physical Science	10	20
Table Reading	7	40
Instrument Comprehension	5	25
Block Counting	4.5	30
Aviation Information	8	20
Total	3 hours, 36.5 minutes	550

VERBAL ANALOGIES

The first part of the Air Force Officer Qualifying Test (AFOQT) begins with a section called "Verbal Analogies." This portion of the test will measure your ability to find correlations between two different words by analyzing definitions or context. You will be given a word in a sequence, and you must choose the best word choice to complete the given sequence. Having a good grasp of vocabulary is vital, but this is also a test of reasoning and critical thinking. The Air Force wants to know that their officers are able to effectively think and communicate.

How The Questions Are Structured

In the Verbal Analogies section, you are given two words that provide a framework for the analogy. The third word then must be paired with the fourth word to parallel the first two words. Think about it like this: A is to B as C is to D.

Here is an example of a verbal analogy:

Tiger (A) is to animal (B) as yellow (C) is to color (D).

In the above example, you see four main words (A, B, C, and D).

Tiger is the first word (A).

Animal is the second word (B).

Yellow is the third word (C).

Color is the fourth word (D).

For verbal analogy questions, you will have to choose the fourth word to complete the sequence. The first two words set up the analogy. In the above example, tiger is a type of animal. Animal is a general category. Yellow is the third word that needs to be paired with the correct category. Yellow is a type of color, such as tiger is a type of animal. Since animal and color are both categories, color is the right word choice.

Determining the Best Answer

Below is an example of how the question is structured on the test.

Joy is to happy as sorrow is to

- A. Love
- B. Truth
- C. Sad
- D. Emotion
- E. Fear

On the test, you will be given several options. It will be multiple choice like in the example above. In the above instance, you are given a pair of words: joy and happy. One synonym for joy is happiness, and sorrow means great sadness. When we look through the possible choices (A. Love, B. Truth, C. Sad, D. Emotion, E. Fear), we can see that the word "sad" would finish the sequence in the most appropriate way.

One way to approach this question is by using a process of elimination. Love (A) does not correlate with sorrow in the way joy does with happy. Neither does truth (B). Sadness is a synonym for sorrow, so sad (C) would be a good choice. At this point we still want to rule out the other options. Sorrow is a type of emotion (D), but emotion is too general of a category. Remember, we are trying to choose something that mirrors the correlation between joy and happy. Fear (E) is not tied to sorrow in the same way sad is. Sad (C) becomes the clear choice when compared with the other options. It is an adjective and synonym of the noun sorrow in the same way that the noun joy is to the adjective happy.

Keep in mind you only have about 19 seconds to evaluate the question, answer it, and move on to the next question. You have 25 questions to answer in 8 minutes. This process of determining the right word to complete the sequence needs to happen very quickly. We will cover other types of word analogies in this chapter, but first let us explore one more type of question structure that you'll encounter.

We have already talked about the "A is to B as C is to ?" structure. You will also see "A is to B as ? is to ?." Check out the following question:

Golf is to sport as:

   A. Jump is to car
   B. Trick is to magician
   C. Pear is to fruit
   D. Carrot is to radish
   E. Monkey is to banana

In the above example, you are given two words that frame the analogy. Golf is to sport as ? is to ?. Let's look at the possible options. Golf is a noun and sport is a noun. Golf is a type of sport. In option A, the word jump is a verb and car is a noun. Jump is something you do to a car. In option B, trick is a noun and it is something that a magician (noun) does. In option C, pear is a noun and fruit is a noun. Pear is a type of fruit. In option D, carrot and radish are both vegetables and both nouns. Option E presents two nouns, monkey and banana. Monkeys eat bananas. Since golf is a type of sport and pear is a type of fruit, the most obvious answer here would be "C. Pear is to fruit."

When completing these types of questions, think of yourself as an investigator. You are not only looking for word definitions, but you are looking for the connection. The Air Force is testing your aptitude for communication. They are testing your effectiveness at connecting the dots and linking different pieces of information. This is not simply a test to know how much of the dictionary you have read. This is a test to see if you can find the relation between two words.

Look for what the two words have in common. Look for how they are related. If you find the common ground between two words, you have found the relationship they have. Once you know the relationship they have, you can look for the second set of words that will replicate that relationship. You must look for clues as you read the first two words to shed light on the second pair of words. For example, take a look at this next analogy:

Cog is to wheel as:

   A. Paw is to cat
   B. Juice is to thirst
   C. Tuck is to roll
   D. Jump is to sit
   E. Peach is to cream

First, look at the first two words and find the common ground. What relationship do they have? Once you see the relationship, which option replicates that relationship? This is a part of whole analogy. Cog is a part of a wheel. The answer would be A, because a paw is part of a cat. The relationship of "cog to wheel" is replicated by "paw to cat." Juice to thirst does not fit. Tuck and roll do not mimic the cog to wheel relationship. Option D clearly does not. And finally, peach to cream is not a parallel connection. Answers B through E do not replicate the part of whole relationship. "Paw is to cat" is the correct answer.

Discovering the Word Relationship

Here is a list of some basic connections you will see between two words; however, words can have endless numbers of connections. As you investigate a word analogy, you can keep these possible correlations in the back of your mind:

- Part of whole
- Synonym/Antonym
- Tool/use or object/function
- Category/Pair/Characteristics
- Intensity
- Cause and effect
- Geographic
- Sequence of events
- Provider/provision
- Source-comprised-of

Part of Whole Analogies

- Wing is to bird
- Steering wheel is to car
- Peel is to banana

Like in the "cog to wheel" example, a part of whole analogy will start by pairing a part with a whole. Be careful with these analogies. As you look at the three examples above, all of them are part of whole analogies, but two are most closely related. At first, it seems like all three have the same relationship, which is true; however, birds and bananas are things of nature while a car is a man-made invention. Sometimes subtle differences can be what makes one analogy a better choice. Always keep in mind that there may be more than one relationship. You want to choose the analogies that have the most things in common.

Synonym/Antonym Analogies

Synonym analogies
- Truth is to fact
- Speak is to talk
- Song is to melody

Antonym analogies
- Day is to night
- Courageous is to fearful
- Tall is to short

Synonym and antonym analogies are pretty straightforward. If you come across a pair of words that have a similar meaning, look for an answer with two words of similar meaning. If there is more than one answer with pairs of synonyms, then look for a second connection. Look at the synonym pairs above. Which two have the most in common? Sure, they are all pairs of synonyms, but pay attention to the *parts of speech*. Notice how truth and fact are both nouns. Also, song and melody are both nouns. Speak and talk are both verbs. What makes the word "talk" hard to discern is that it is sometimes a noun (ex: "Let's sit down and have a talk"). What gives us a clue about "talk" being a verb in this instance is that "speak" is a verb. Remember that it is all about word relationships, not just their definitions. Discerning the parts of speech is very difficult when you are not given a full sentence, but looking for the relationship and potential relationship between the two words will help shed light on what part of speech it is.

For antonym analogies, you may see two words that contrast one another. If the words have opposite definitions, it is most likely an antonym analogy. If that is the case, simply look for an answer that has two words with opposite meanings. In the antonyms above, they have a similar relationship in that they are pairs of words with opposite definitions. If I asked you, "Which pair has the most in common?" which two pairs would you choose? Since "courageous is to fearful" and "tall is to short" are both pairs of adjectives, you would want to choose these two pairs as being most similar. Once again, you want to choose the best answer, not just a good answer.

Tool/Use Analogies

- Baseball is to throw
- Phone is to call
- Shelf is to books

In tool/use analogies, you will notice the connection is how something is used. Baseballs are meant to be thrown. Phones are meant to make calls. Shelves are meant to hold books. From the analogies listed above, the two that have the most in common are "baseball is to throw" and "phone is to call." They both share a "noun is to verb" sequence, whereas "shelf is to books" is a "noun to plural noun" sequence.

Category Analogies

- Blue is to color
- Nike is to shoe
- Scotch is to tape

Category analogies have a typical "type to category" sequence. Blue is a type of color. Nike is a type of shoe. Scotch is a type of tape. In the above examples, "Nike is to shoe" and "Scotch is to tape" have the most in common because they are both brands. They are brand types of product categories. "Blue is to color" is the odd man out, even though it is a "type of category" sequence.

Intensity Analogies

- Trickle is to pour
- Lukewarm is to boiling
- Tired is to exhausted

Here are some examples of intensity analogies. The words may have similar meanings, but they are not synonyms. One word is a more extreme version of the other word.

Cause and Effect Analogies

- Careless is to accident
- Careful is to safety
- Heavy rain is to flood

Cause and effect analogies are similar to sequence in that one word typically precedes the other. But the difference here is that one word causes the effect. Accidents are caused by carelessness. Floods could be caused by heavy rain. Being careful could cause safety.

Geographical Analogies

- The Statue of Liberty is to New York
- The Eiffel Tower is to Paris
- Hawaii is to the ocean

Geographical analogies have to do with the location of the two words. The Statue of Liberty is in New York. Paris is a major city in France. Hawaii is surrounded by ocean water. The Eiffel Tower is to Paris as the Statue of Liberty is to New York. They are both major attractions in major cities.

Sequence of Events Analogies

- Fall is to winter
- Spring is to summer
- Baby is to toddler

Sequence of events analogies have to do with the sequence of things. A human is a baby before they are a toddler. Fall comes before winter and spring comes before summer. These two analogies have the most in common because they are both seasons.

Provider/Provision Analogies

- Baker is to bread
- Logger is to timber
- Barista is to coffee

Provider/provision analogies are often vocational, including a person who provides a good or service. In this case, bakers make bread. Loggers cut down trees, called timber. Baristas make coffee. All three of these examples are provider/provision analogies, but the "baker is to bread" and "barista is to coffee" analogies have the most in common. They are both food items that the provider makes.

Source-Comprised-Of Analogies

- Lake is to water
- Shirt is to cotton
- Road is to asphalt

Source-comprised-of analogies have two nouns. One noun is comprised of the other. Lakes are made of water. Shirts are made of cotton. Roads are made of asphalt. One thing is comprised of the other (the source). In the three analogies above, "shirt is to cotton" and "road is to asphalt" are the most similar in that they are both man-made.

Other Analogies You May Encounter

You may encounter all sorts of analogies on the AFOQT. An analogy is simply a comparison between two things. You are inspecting the verbal analogy for a relationship. Once you define the relationship between the initial two words, you'll have a better idea as to what the answer should be. Remember, you are choosing the best choice from the options. Sometimes the answer may not be a perfect parallel analogy, but it is the best option from the multiple choice options.

Quick Tip: Create a Sentence

If you are having trouble finding the relationship between two words, try to create a sentence out of them. For example, if you are given "child is to school," you could create the following sentence: "The child went to school and was in class all day." The sentence can help you visualize the relationship between the two words. Let's take another example from above, "careless is to accident." The sentence you create could be similar to this: "The teenage boy was being careless with his driving, so he got into an accident." This sentence helps you visualize the cause and effect analogy. You may not need to use this tool, but some people have to visualize two words together by creating a sentence. Do not use this tool if you don't need it, as it will take more time. Time is a precious commodity on the AFOQT.

Before You Start the Practice Questions

On the next page, you will begin a Verbal Analogies practice test. Set a timer for 8 minutes before you start this practice test. Giving yourself 8 minutes will give you an authentic feel for how long you have to finish this portion of the AFOQT. Like the official test, this practice test has exactly 25 items. Set your timer and begin.

Verbal Analogies Practice Test Questions

(8 minutes)

1. Run is to jump as
    a. Sit is to stand
    b. Drink is to sip
    c. Swing is to swung
    d. Poke is to hit
    e. Speak is to say

2. Firm is to solid as
    a. Grab is to handle
    b. Work is to bend
    c. Wrench is to screwdriver
    d. Marathon is to running
    e. Dry is to parched

3. String is to guitar as
    a. Pork is to meat
    b. Tie is to suit
    c. Apple is to orange
    d. Water is to river
    e. Hose is to fire

4. Table is to food as
    a. Sun is to rise
    b. Ocean is to boat
    c. Butter is to bread
    d. Pepper is to salt
    e. Juice is to fruit

5. Hertz is to frequency as Kelvin is
    a. Alkaline
    b. Depth
    c. Temperature
    d. Amplitude
    e. Circuit

6. Catastrophic is to ruinous as insipid is to
    a. Strenuous
    b. Bland
    c. Cyclical
    d. Torrent
    e. Chill

7. Arduous is to facile as grandiose is to
    a. Mooring
    b. Amplitude
    c. Gradual
    d. Mediocre
    e. Fortitude

8. Barber is to hair as optometrist is to
    a. Eyes
    b. Feet
    c. Window
    d. Hardware
    e. Math

9. Day is to early as night is to
    a. Late
    b. Dinner
    c. Dusk
    d. Dawn
    e. Evening

10. Box is to cube as ball is to
    a. Circle
    b. Prism
    c. Sphere
    d. Round
    e. Shape

11. Punctual is to time as fluent is to
    a. River
    b. Water
    c. Language
    d. Speaking
    e. Pipe

12. Erroneous is to inaccurate as horrendous is to
    a. Harrowing
    b. Indecent
    c. Modest
    d. Tepid
    e. Visceral

13. Walk is to strolling as
    a. Cigar is to smoking
    b. Drive is to jaunting
    c. Truth is to telling
    d. Water is to flowing
    e. Wonder is to thinking

14. Bread is to wheat as glass is to
    a. Clear
    b. Sand
    c. Crystal
    d. Field
    e. Water

15. Gander is to geese as
    a. Pork is to pig
    b. Cookie is to crumb
    c. Bird is to fly
    d. Hunt is to game
    e. Vixen is to fox

16. Crime is to criminal as clean is to
    a. Bleach
    b. Brush
    c. Wash
    d. Bathroom
    e. Janitor

17. Push is to pull as
    a. Chocolate is to vanilla
    b. Truth is to tell
    c. Sit is to fall
    d. Work is to march
    e. Answer is to question

18. Shovel is to dig as
    a. Chair is to sit
    b. Kite is to fly
    c. Poems are to read
    d. Books are to write
    e. Noise is to hush

19. Rose is to flower as red is to
    a. Song
    b. Dress
    c. Blood
    d. Color
    e. Mercury

20. Stand is to fall as swim is to
    a. Water
    b. River
    c. Wet
    d. Pool
    e. Sink

21. Ignite is to fire as hurt is to
    a. Pain
    b. Heal
    c. Hospital
    d. Emergency
    e. Match

22. Los Angeles is to California as
    a. State is to country
    b. Milk is to jug
    c. Country is to city
    d. Road is to highway
    e. Exit is to freeway

23. Equivocal is to ambiguous as
    a. Smart is to ingenious
    b. Run is to exercise
    c. Talk is to communicate
    d. Eat is to food
    e. Work is to play

24. Carrot is to broccoli as
    a. Bark is to dog
    b. Apple is to fruit
    c. Pigeon is to dove
    d. Whale is to ocean
    e. Rain is to cloud

25. Whisper is to shout as
    a. Sun is to rain
    b. Day is to night
    c. Walk is to run
    d. Rise is to fall
    e. Swim is to dive

Answer Guide to Verbal Analogies Practice Test

1. **A.** This is a state of action analogy. Run and jump are two different verbs that a person can do. Sit and stand are two verbs a person can do. B is two different levels of one action, sip or drink. C is one action in the present and past tense. Speak and say are synonyms. Answer A is the more similar relationship.
2. **E.** A is a verb to a noun. B doesn't seem to have any similar connection. C is unrelated. D is a noun to a verb. Dry and parched are synonyms, much like firm and solid.
3. **B.** This is a part of whole analogy. Pork is a type of meat. Tie is a part of a suit. Apple and oranges are types of fruit. Water is what a river is made from. Hose and fire are not related in the same way as string is to guitar.
4. **B.** This is an object/function analogy. Suns rise. Boats are on oceans, much like food is on the table. C is tricky because at first it appears that butter is on bread, much like food is on the table. You must pay attention to the order of words. Table is to food as butter is to bread. Butter goes on bread, but a table does not go on food. The order of words is what rules out this analogy. Pepper and salt are a pair. Juice is derived from fruit.
5. **C.** Hertz is a measurement of frequency. Kelvin is a measurement of temperature.
6. **B.** Catastrophic is a synonym of ruinous. Insipid is a synonym of bland.
7. **D.** Arduous and facile are antonyms. Grandiose and mediocre are antonyms.
8. **A.** This is a partial provider/provision analogy. Barbers work with hair. Optometrists work with eyes. People go to barbers to have them work on their hair. People visit optometrists for them to work on their eyes.
9. **A.** Day and night are antonyms. Early and late are antonyms. Day corresponds with early in a parallel way that night corresponds with late. This a partial sequence of events analogy as well as a synonym/antonym analogy.
10. **C.** This is a characteristic analogy. A box is a cube. A ball is a sphere. Yes, a ball is round, a circle, and a shape. But the 3-dimensional shape of a box is a cube, and the 3-dimensional shape of a ball is a sphere. C is the parallel analogy.
11. **C.** Punctual is an adjective of time as fluent is an adjective of language. Speaking is a verb whereas time is a noun. Language is a noun, so the relationship between fluent and language is parallel to the relationship between punctual and time.
12. **A.** Erroneous and inaccurate are synonyms, much like horrendous and harrowing are synonyms.
13. **B.** Strolling is a way someone may walk. Jaunting is a way someone may drive. The two phrases are most similar compared with the other options. In E, everyone who wonders thinks, but not everyone who thinks wonders. Strolling is a type of walking, but thinking is not a type of wondering. For C and D, water and truth are

both nouns. Walk can be a noun, but it has the potential to be a verb, much like the word "drive." A is an object/function analogy.

14. **B**. This is a source-comprised-of analogy. Bread is made from wheat. Glass is made from sand. Technically, bread could be made from rice, oats, or something else. Sometimes you need to "read between the lines" and think like the test maker. Bread is usually made of wheat.

15. **E**. Gander is a male goose. Vixen is a female fox. Gander is to geese as vixen is to fox. Geese is plural and fox is singular, so sometimes these analogies don't always parallel exactly, but it is a *better* choice than the others. Pork is what a pig is made of. Crumb is a part of a cookie. Fly is the action of a bird. Game is something one might hunt. Nothing mimics the relationship of "gander is to geese" as much as "vixen is to fox."

16. **E**. This is a provider/provision analogy. Crime is something criminals do. Clean is something janitors do. Bleach and brush may clean, but they are not human like a janitor. Criminals are human.

17. **E**. Push is an antonym of pull as answer is to question. Sit and fall are not antonyms (maybe sit/stand or fall/rise). Chocolate and vanilla are pairs, not antonyms. If someone argued that chocolate is an antonym of vanilla, I would say that push/pull and answer/question can be verbs. Chocolate and vanilla cannot be verbs. E is the best choice.

18. **B**. This is a tool/use analogy or an object/function analogy. Shovel is to dig as kite is to fly. Chairs do not sit. They are sat on. Poems do not read. They are read. Books do not write. They are written. Kites do the flying like shovels do the digging. Also, they both need human guidance.

19. **D** Roses are a type of flower, and red is a type of color. This is a category analogy.

20. **E**. Standing is a gravitational resistance to falling as swimming is a gravitational resistance to sinking. These are the most parallel analogies.

21. **A**. This is a cause and effect analogy. Fire is an effect caused by igniting. Pain is an effect caused by hurting.

22. **A**. This is a geographical analogy. Cities are within states geographically. States are within countries geographically. Los Angeles is within California geographically as a state is within a country geographically. Milk is inside a jug, but it is not as similar geographically. Remember, you are trying to choose the *best* choice.

23. **A**. Equivocal and ambiguous are synonyms like smart and ingenious. Talk and communicate can be synonyms, but they are verbs instead of adjectives. Run is a type of exercise, but it is not a synonym for exercise.

24. **C**. This is a pair analogy. Carrots and broccoli are both vegetables. Pigeons and doves are both birds. Bark is something a dog does. Apple is a type of fruit. Whale is in an ocean. Rain comes from a cloud.

25. C. This is an intensity analogy. Running is a more intense form of walking. Sun is the opposite of rain. Day is opposite of night. Rise is opposite of fall. Swim and dive are two actions, not necessarily a more intense version of each other. Shout is an intense version of whispering like running is an intense version of walking.

ARITHMETIC REASONING

Word problems make up most of the Arithmetic Reasoning section of the AFOQT. To solve the problems, you will need to use reasoning and mathematics. You will be given a list of possible answers to solve everyday situations. Fractions, ratios, averages, percentages, and rates are all things you will need to know to do well during this part of the test. Later in this chapter, you will have the chance to complete some practice problems that will help prepare you for the real AFOQT.

Preparation Strategy

In this chapter, we will cover the necessary math skills you will need to solve each word problem. These are basic arithmetic elements that you may be proficient in already. You need to be able to read each word problem and know which tools to apply to get the correct answer.

Do not simply read through the problems without practicing them. Math requires practice to be competent. Make sure to do the practice questions and then compare your answers with the provided correct answers. Sometimes students read through the examples and think that they can do the problems. What matters is that you can do the problems on your own, not simply read through examples. Note that many times, there are multiple approaches to the same problem. As long as you get the correct response in a timely manner, don't worry if you took a slightly different approach than in the answer guide provided.

Classification of Numbers

Numbers are the substance that math is made of. Different classifications of numbers are used to communicate different math concepts. The following definitions of numbers will help navigate the language used in various operations of math.

Whole numbers are numbers with no fractions or decimals. 3 is a whole number. ¾ (or .75) or 3.75 are not whole numbers. Also, whole numbers do not include negative numbers.

Integers are any positive numbers or negative numbers without decimals, including zero. Decimals (.75 or 3.75), fractions (¾), or mixed numbers (3¾) are not integers. The difference between whole numbers and integers is that an integer can be negative (-1, -2, etc...).

Factors are the numbers you multiply together to get another number. For example, 1, 2, 4, and 8 are factors of 8 because $1 \times 8 = 8, 2 \times 4 = 8$. 1, 2, 4, and 8 can be multiplied together in different ways to become 8, which makes them factors of 8.

Remainders are what is left over after doing division. For example, dividing 3 by 2 would leave a remainder of 1. No remainder would exist if we divided 4 by 2. We will cover this in more detail later on in this chapter.

Even number: any integer that can be divided by 2 without leaving a remainder. Examples of even numbers are 2, 4, 6, etc...

Odd number: any integer that *cannot* be divided evenly by 2. Examples of odd numbers are 1, 3, 5, etc...

Prime number: any whole number greater than 1 that has only two factors: itself and 1. It can only be divided evenly by itself and 1. Examples of prime numbers are 2, 3, 5, 7. 4 is not a prime number because it can be divided evenly by 2.

Composite number: any whole number greater than 1 that has more than two factors. Essentially, a composite number is any number that is not a prime number. For example, 6 is a composite number because it can be divided evenly by 1, 2, 3, and 6 (its factors are 1, 2, 3, and 6).

Decimal number: any number that uses a decimal point to show part of the number that is less than the next whole number. For example, 2.33.

Decimal point: a symbol used to separate the ones from the tenths place or dollars from cents in currency.

Decimal place: the position of a number to the right of a decimal point. For example, in the decimal 1.345, the 3 is in the tenths place, the 4 is in the hundredths place, and the 5 is in the thousandths place.

Digit: any numeral 0 through 9.

Rational number: A number which can be expressed as $\frac{a}{b}$ where *a* and *b* are *integers* is said to be rational. The rational numbers include all the integers, since any integer can be written as a fraction with a denominator of 1.

Irrational number: A real number that is not rational is, by definition, irrational. Some common examples of irrational numbers include $\pi, e,$ and $\sqrt{2}$. If written as a decimal, irrational numbers never terminate and often keep repeating. For example, 2.66666.... It may surprise you to know that *most* of the numbers that exist are irrational. For example, there are an infinite number of numbers between 1 and 1.000000001, and most of those are irrational.

Real numbers: Real numbers are all of the rationals (and so also all of the integers, fractions, and decimals) together with the irrationals. If you were to throw a dart at a number line, with a very, very accurate measuring system, your dart would land on a real number. In fact, the only numbers that aren't real are *imaginary* – and you won't encounter those anywhere on a number line.

Place Value

When referring to specific digits in a number, it is important to understand what each place value is called. Below are the place values of each digit in the number 23,154.987:

2: ten-thousands
3: thousands
1: hundreds
5: tens
4: ones
9: tenths
8: hundredths
7: thousandths

Primary Arithmetic Operations

Four foundational arithmetic operations exist when working with numbers: addition, subtraction, multiplication, and division.

- Addition takes different numbers and combines them into a total, or sum. Sum means total when combining different collections into one. If there are 4 things in one collection and 5 things in the other, then after combining them, there is a total of $4 + 5 = 9$. Order doesn't matter when adding numbers. The equation could also look like $5 + 4 = 9$.
- Subtraction is the opposite, or inverse, operation of addition. Addition combines numbers together, but subtraction takes one quantity away from the other. For example, if there are 12 eggs and 3 are removed, that gives

us the equation $12 - 3 = 9$ eggs remaining. In subtraction, the order *does* matter because we need to know the amount that is being taken away and from which number it is being taken from.
- Multiplication can be thought of as repeated addition. For example, 5×3 can be thought of as adding 5 sets of 3's or 3 sets of 5's. Another way to think about it is 3 sets containing 5 items, totaling 15. If you are more of a geometrical thinker, imagine a rectangle 5 tiles wide and 3 tiles long. You have three rows of 5 tiles or you have 5 rows of 3 tiles. Either way, you will count 15 tiles. In multiplication, order of operation does not matter. 3×5 is the same as 5×3; both equal 15.
- Division is the opposite, or inverse, of multiplication. You take one quantity and divide it into sets that are the size of the second quantity. If there are 12 donuts to be given to 3 people, then each person gets $12 \div 3 = 4$ donuts. Order matters with division. $12 \div 3$ is different than $3 \div 12$.

Parentheses

Parentheses are used to designate which operation should be done first. For example, $5 - (3 - 1) = 3$; since $(3 - 1)$ is in parentheses, it is done first. $3 - 1 = 2$. Then we subtract 2 from 5: 5-2=3. If we ignored the parentheses, the answer would be $5 - 3 - 1 = 1$. The right answer is 3, but if we ignore the parentheses, the answer is 1, which would be the wrong answer.

Addition

Addition is denoted with the + symbol.

Addition follows the *commutative property*. This means that numbers can be added in different orders with the same result. For example: $5 = 3 + 2 = 2 + 3 = 5$. The formula for the commutative property of addition is $a + b = b + a$.

Addition also follows the *associative property*. This means that the grouping of numbers or the placement of parentheses does not matter.
For example: $15 = (5 + 3) + 7 = 5 + (3 + 7) = 15$. The formula for the associative property of addition is $a + (b + c) = (a + b) + c$.

Subtraction

Subtraction is denoted by the − symbol. Subtraction involves taking one number away from another. The result is referred to as the *difference*.

Subtraction follows neither the commutative nor associative properties. In other words, the order of numbers and the placement of parentheses is important, as it affects the outcome of an equation. Look at the four different examples below:

$$10 = 12 - (6 - 4)$$
$$2 = (12 - 6) - 4$$
$$5 = 10 - 5$$
$$-5 = 5 - 10$$

In the above examples, remember to start with the numbers in parentheses first before moving on to the rest of the equation.

When working with larger numbers, it is helpful to regroup the numbers. Take for instance the problem below:

$$642 - 85$$

Let's write this subtraction problem vertically, and group the numbers by column:

$$\begin{array}{r} 642 \\ -\ 85 \\ \hline \end{array}$$

The ones and tens column of 85 (8 is in the tens column and 5 is in the ones column) is larger than the ones and tens column of 642 (6 is in the hundreds column, 4 is in the tens column, and 2 is in the ones column).

To simplify this formula, *borrow* from the hundreds column and the tens column.

When borrowing from a column, subtracting 1 from the lender column will add 10 to the borrow column.

First, borrow from the hundreds and add 10 to the tens...

$$\begin{array}{r} ^{5}\!\!\!\!\! \\ \cancel{6}\ 14\ 2 \\ -\ \ \ 8\ 5 \\ \hline \end{array}$$

Then, borrow from the tens and add 10 to the ones.

$$\begin{array}{r} 5\ \ 13 \\ \cancel{6}\ \cancel{14}\ 12 \\ -\ \ 8\ \ 5 \\ \hline \end{array}$$

Then, you can subtract down the columns.

$$\begin{array}{r} 5\ \ 13\ \ 12 \\ -\ \ \ \ 8\ \ \ 5 \\ \hline 5\ \ \ 5\ \ \ 7 \end{array}$$

So, $642 - 85 = 557$.

As you can see, after borrowing, the digits in the top row will be larger than the digits in the bottom row before proceeding. Honing this simple technique can help with subtraction involving larger numbers!

Multiplication

Multiplication is denoted by the $(\cdot)$, $(\times)$, or $(*)$ symbols. Also, multiplication is denoted by the use of parentheses next to a number. For instance:
$$5(4) = 5 \times 4 = 5 * 4 = 5 \cdot 4 = 20.$$

The numbers being multiplied together are referred to as *factors,* and the answer or result is the *product.* In the equation $5 \times 3 = 15$, 5 and 3 are factors. 15 is the product.

Like addition, multiplication follows the commutative and associative properties:

$$84 = 12 \times 7 = 7 \times 12 = 84$$
$$126 = 3 \times (7 \times 6) = (3 \times 7) \times 6 = 126$$

Multiplication follows the *distributive* property. The formula for the distributive property of multiplication is $a \times (b + c) = ab + ac$. In other words, when the sum of two numbers b and c is multiplied by a single number a, you can "distribute" the a by multiplying each b and c by a, and then adding those products. This will give you the same answer as if you added b and c, and then multiplied that sum by a.

It is much easier to understand this by examining the following examples:

$$50 = 5 \times 10 = 5(6+4) = (5 \times 6) + (5 \times 4) = 30 + 20 = 50$$
$$117 = 9(7+6) = (9 \times 7) + (9 \times 6) = 63 + 54 = 117$$

Sometimes, you will need to multiply large numbers. Thankfully, this can be done quickly, thanks to an algorithm which is similar to the one we used for subtraction, above. Consider the following problem:

$$284 \times 63$$

Let's write this multiplication problem vertically and group the numbers by column:

$$\begin{array}{r} 2\,8\,4 \\ \times\,6\,3 \\ \hline \end{array}$$

When you set up a multiplication problem like this, begin by writing the larger number on top of the smaller one, like this. Then, begin by looking at the number at the bottom of the ones column. In this case, it's a 3.

This 3 is going to multiply *each* of the numbers above it, beginning with the ones (a 4), then the tens (an 8), and then the hundreds (a 2). Each number in the bottom will multiply each of the numbers above it in turn. This will become more clear as we proceed. Let's begin by multiplying down the ones column.

$$\begin{array}{r} 2\,8\,4 \\ \times\,6\,3 \\ \hline 12 \end{array}$$

We got a two-digit number, but only one of those fits in the ones column. So, keep the 2, and "carry" the 1 to the top of the next column, in this case the tens.

$$\begin{array}{r} ^{+1} \\ 2\,8\,4 \\ \times\,6\,3 \\ \hline 2 \end{array}$$

Next, multiply the 3 by the top number in the next column (the 8). 3 times 8 is 24, but don't forget we carried a one, so we have to add it to the product after we've done our multiplication. So, we have $8 \times 3 + 1 = 25$.

$$\begin{array}{r} {\scriptstyle +1} \\ 2\ 8{\times}3\ 4 \\ \times\ 6\ 3 \\ \hline 25\ 2 \end{array}$$

Once again, we'll have to carry our second digit.

$$\begin{array}{r} {\scriptstyle +2} \\ 2\ 8\ 4 \\ \times\ 6\ 3 \\ \hline \ 5\ 2 \end{array}$$

And repeat: $2 \times 3 + 2 = 6 + 2 = 8$.

$$\begin{array}{r} {\scriptstyle +2} \\ 2{\times}38\ 4 \\ \times\ 6\ 3 \\ \hline 8\ 5\ 2 \end{array}$$

That takes care of the 3, but now we have to do the same thing in the tens column, and the 6 will be multiplied by each of the numbers up top, carrying extra digits as we go. First, we add a zero to the ones column to signify that we're working on the tens now.

$$\begin{array}{r} 2\ 8\ 4 \\ \times\ 6\ 3 \\ \hline 8\ 5\ 2 \\ 0 \end{array}$$

Then, we'll multiply the 6 by each of the numbers up top and bring down those results. First, $6 \times 4 = 24$, and we will need to carry the 2.

$$\begin{array}{r} {\scriptstyle +2} \\ 2\ 8\ 4 \\ \times\ 6\ 3 \\ \hline 8\ 5\ 2 \\ 4\ 0 \end{array}$$

Then, $6 \times 8 + 2 = 48 + 2 = 50$. Likewise, we'll carry that 5.

```
   +5
   2  8  4
 x    6  3
   8  5  2
   0  4  0
```

And finally, $6 \times 2 + 5 = 12 + 5 = 17$. Since this is our last column, the extra digit can just fall into place.

```
      2  8  4
    x    6  3
      8  5  2
  17  0  4  0
```

From here, just add the two numbers, carrying digits as necessary (but in this example, we don't need to).

```
       2  8  4
     x    6  3
       8  5  2
  +17  0  4  0
   17  8  9  2
```

So, $284 \times 63 = 17{,}892$.

Division

Division is denoted by the $(\div)$ *and* $(/)$ symbols. Much like addition and subtraction are opposite of one another, division is the inverse of multiplication.

Sometimes division can be denoted as one number over another one; in other words, as a fraction such as $\frac{5}{8}$. In this situation, the number on top is called the *numerator,* and the number on the bottom is called the *denominator*. You may be accustomed to seeing fractions with a smaller number on top and a larger number on bottom, but something like $\frac{24}{8}$ is also valid. This fraction could be interpreted as $24 \div 8$. When written in this way, the number before the division symbol (24) is called the *dividend* and the number after the division symbol (8) is called the *divisor*. So, numerators are dividends and denominators are divisors.

Like subtraction, division does not follow the commutative property. It matters what number comes before or after the division symbol. It also does not follow the associative or distributive properties for the same reason. Here are some examples:

Division is not commutative; that is, $8 \div 4 \neq 4 \div 8$:

$$8 \div 4 = \frac{8}{4} = 2.$$
$$4 \div 8 = \frac{4}{8} = \frac{1}{2}.$$

Division is not associative; that is, $(48 \div 2) \div 6 \neq 48 \div (2 \div 6)$:
$$(48 \div 2) \div 6 = 24 \div 6 = 4.$$
$$48 \div (2 \div 6) = 48 \div \frac{2}{6} = 144.$$

Finally, division is not distributive; that is, $(30 \div 2) + (30 \div 3) \neq 30 \div (2 + 3)$:

$$(30 \div 2) + (30 \div 3) = 15 + 10 = 25.$$
$$30 \div (2 + 3) = 30 \div 5 = 6.$$

Remainders

In the above examples, we have largely been working with divisors that are factors of their dividends. For example, $30 \div 3 = 10$ because $10 \times 3 = 30$. However, often in division we will want to divide numbers that are not factors of their dividends, such as $30 \div 7$. In this case, we wish to know what we must multiply 7 by to be as close to 30 as possible without going over, and how much remains. The value left over, the value which remains, is called the *remainder.* In this case, $7 \times 4 = 28$, and there are 2 left over to get to 30, so $30 \div 7 = 4$ with a remainder of 2.

Since division in the case of remainders can be a little less clear, often long division is a good way to approach division. It enables us to divide whole numbers into larger numbers more easily and will naturally lead to a remainder, if one exists.

Long Division

Long division is a method of division that develops an algorithm similar to what we have used when adding, subtracting, and multiplying larger numbers. Unlike those methods, we don't write the numbers on top of each other in long division. Instead, we use the long division symbol:

$$415 \div 8 = 8\overline{)415}$$

divisor — dividend

Once written this way, long division is a matter of dividing the divisor into each digit of the dividend individually, beginning at the left, and carrying over the value that remains in each column.

First, we consider $4 \div 8$.

$$\begin{array}{r} 0 \\ 8\overline{)415} \end{array}$$

Since 8×1 is too large, we must use 8×0. Place the result below the first column, subtract, and drop down the next digit.

$$\begin{array}{r} 0 \\ 8\overline{)415} \\ -0 \\ \hline 41 \end{array}$$

Next, we will consider $41 \div 8$, which is 5 (with a little left over).

Write a 5 up top. Since $5 \times 8 = 40$, we subtract 40 from the 41 we have, leaving a 1, and drop down the next digit.

$$\begin{array}{r} 05 \\ 8\overline{)415} \\ -0 \\ \hline 41 \\ -40 \\ \hline 15 \end{array}$$

Now, 15/8 is 1, with a little left over.

```
  051
8)415
  -0
  41
 -40
  15
  -8
   7
```

So, 415 divided by 8 is 51, with a remainder of 7.

```
  051 r7
8)415
  -0
  41
 -40
  15
  -8
   7
```

The number you get during division that isn't the remainder is called the *quotient*. You can check your work, as well as illustrate the relationship between division and multiplication, by multiplying the divisor and quotient and adding the remainder.

Divisor x Quotient + Remainder = Dividend

$$8 \times 51 + 7 = 408 + 7 = 415$$

If you end up with a remainder of zero, your divisor is a *factor* of the dividend. For example, if 408 was our dividend, not 415, then the remainder would come out to be zero, making 51 and 8 factors of 408.

Using long division with remainders is a good way to divide large numbers.

Fractions

A fraction is expressed as one integer on top of another with a dividing line between them ($\frac{x}{y}$). It can be thought of as x divided by y. It can also be thought of as x number of parts out of y number of equal parts.

The top number is called the *numerator*. The bottom number is called the *denominator*. The denominator cannot be zero (this is referred to as *undefined*).

Simplifying Fractions

Fractions can be simplified, or reduced, by dividing the numerator and the denominator by the same number. Fractions that have the same value but are expressed differently are known as equivalent fractions. For example, $\frac{2}{4}, \frac{4}{8}$, and $\frac{8}{16}$ are all equivalent fractions. By dividing the numerator and the denominator in each fraction by a common factor, they can all be reduced or simplified, to $\frac{1}{2}$.

Common Denominators

Sometimes fractions are manipulated so that they have the same denominator. This is known as finding a *common denominator*. The number that is chosen to be the common denominator should be the *least common multiple* of the two original denominators. Take for example $\frac{2}{5}$ and $\frac{2}{3}$. The least common multiple of 5 and 3 is 15; hence $\frac{2}{5} = \frac{6}{15}$ and $\frac{2}{3} = \frac{10}{15}$. Finding a common denominator can make addition and subtraction of fractions easier.

Mixed Numbers

When the numerator is less than the denominator in a fraction, it is known as a *proper fraction*. An *improper fraction* is one in which the numerator is greater than the denominator. Proper fractions have values less than one, and improper fractions have values greater than one.

A mixed number is a number that contains an integer and a fraction. Because improper fractions have a value greater than one, any improper fraction can be rewritten as a mixed number. For example, $\frac{9}{4} = \frac{8}{4} + \frac{1}{4} = 2 + \frac{1}{4} = 2\frac{1}{4}$. Also, any mixed number can be rewritten as an improper fraction. For example, $1\frac{2}{5} = 1 + \frac{2}{5} = \frac{5}{5} + \frac{2}{5} = \frac{7}{5}$.

Adding and Subtracting Fractions

When two fractions have the same denominator, they can be added or subtracted by adding or subtracting the numerators and retaining the same denominator. For example,

$\frac{1}{3} + \frac{2}{9} = \frac{3}{9} + \frac{2}{9} = \frac{5}{9}$. If the fractions do not have the same denominator, they must be manipulated to achieve a common denominator before addition or subtraction takes place.

Multiplying Fractions

Fractions can be multiplied by multiplying the two numerators to find the new numerator and multiplying the two denominators to find the new denominator. For example, $\frac{1}{2} \times \frac{3}{5} = \frac{3}{10}$.

Dividing Fractions

Two fractions can be divided by turning the second fraction upside down, converting the division sign to a multiplication sign, and proceeding with multiplication. For example, $\frac{1}{3} \div \frac{7}{8} = \frac{1}{3} \times \frac{8}{7} = \frac{8}{21}$.

Decimals

Adding and Subtracting Decimals

When adding or subtracting decimals, it is important to align the decimals properly. Adding or subtracting decimals is just like adding or subtracting whole numbers. For example, $2.6 + 3 = 5.6$. If the decimals are not aligned properly, the answer may be written as 2.9, which would be incorrect and hurt one's test score. Aligning decimals vertically can help:

  2.6
+3.0
  5.6

Subtraction of decimals follows the same set of principles. Aligning the numbers vertically can help avoid incorrect answers:

  5.6
-2.0
  3.6

Multiplying Decimals

Multiplication can be a little more challenging when multiplying numbers with decimals. One simple way to approach these problems is to ignore the decimals until the end and multiply the numbers as if they were whole numbers. After multiplying the factors, place the decimal in the product. The decimal placement is determined by adding up the total number of decimal places in the factors. To understand how this works, consider the examples below:

Example 1	Example 2	Example 3
3.5 × 2.2 = 7.70	1.2 × 1.7 = 2.04	0.8 × 2 = 1.6

Example 1: Ignore the decimal and multiply 35 by 22 to end up with 770 as the product. Next, add up the number of digits that follow a decimal in the factors. 3.5 and 2.2 both have decimals and one digit follows each decimal (5 and 2) for a total of **two** digits. In the product, we place a decimal **two** places to the left: 7.70. Since the zero is now unnecessary, the answer (or product) is 7.7.

Example 2: Ignore the decimal and multiply 12 by 17 for a product of 204. Count the number of digits in the factors that follow decimals. 1.2 and 1.7 have one digit each (2 and 7) that follow decimals for a total of **two** digits. Place the decimal **two** places to the left for a product of 2.04.

Example 3: Ignore the decimal and multiply 8 by 2 for a product of 16. Count the digits in the factors that follow decimals. 0.8 has one digit that follows a decimal (the 8) and 2 has no digits that follow a decimal (it's a whole number with no decimal), which gives us a total of **one** digit. Place the decimal to the left **one** place for a product of 1.6.

Dividing Decimals

To divide decimal numbers, multiply the *dividend* and the *divisor* by 10 until there are no decimals, and then solve the problem. Look at the example below:

$$4.2 \div .7 = 42 \div 7 = 6$$
$$.244 \div .061 = 244 \div 61 = 4$$

The process of long division for large numbers described earlier in the chapter applies here as well!

Percentages

Percent means "per hundred." Percentages can be thought of as fractions with 100 being the whole. Percentages can be expressed as fractions. Simply divide the percentage by 100 and reduce the fraction to its simplest terms. For example, $56\% = \frac{56}{100} = \frac{14}{25}$; $20\% = \frac{20}{100} = \frac{1}{5}$.

Additionally, fractions can be expressed as percentages. Simply manipulate the fraction so that it has a denominator of 100. This can be done by multiplying both the numerator and the denominator by the same number in order for the denominator to be equal to 100. For example, $\frac{2}{5} = \frac{40}{100} = 40\%$; $\frac{9}{10} = \frac{90}{100} = 90\%$.

Converting Percentages, Decimals, Fractions

Sometimes problems require the conversion of percentages, decimals, or fractions. This section covers the process of these conversions.

Converting Percentages to Decimals and Fractions

When converting percentages to decimals, move the decimal point two places to the left. For example, 24% = .24; 2.5% = .025; 500% = 5.00; .22% = .0022.

When converting percentages to fractions, divide by 100 and reduce the fraction into the simplest terms possible. For example, $20\% = \frac{20}{100} = \frac{1}{5}$; $15\% = \frac{15}{100} = \frac{3}{20}$; $7\% = \frac{7}{100}$.

Converting Fractions to Percentages and Decimals

When converting fractions to percentages, multiply the numerator and denominator by a factor that will equal a product of 100 in the denominator. For example, $\frac{3}{5} = \frac{3 \times 20}{5 \times 20} = \frac{60}{100} = 60\%$; $\frac{1}{4} = \frac{1 \times 25}{4 \times 25} = \frac{25}{100} = 25\%$.

When converting fractions to decimals, multiply the numerator and denominator so that the fraction has a denominator of 100. For example, $\frac{7}{10} = \frac{7 \times 10}{10 \times 10} = \frac{70}{100} = .70 = .7$; $\frac{1}{25} = \frac{1 \times 4}{25 \times 4} = \frac{4}{100} = .04$.

Converting Decimals to Percentages and Fractions

When converting decimals to percentages, multiply the decimal by 100 or just move the decimal point two places to the right. For example, $.172 \times 100 = 17.2\%$; $.7 \times 100 = 70\%$; $.002 \times 100 = .2\%$.

When converting decimals to fractions, multiply by $\frac{10}{10}$ until the decimal is gone and then reduce the fraction into its simplest possible terms. For example, $.12 = .12 \times \frac{100}{100} = \frac{12}{100} = \frac{3}{25}$; $.0025 = .0025 \times \frac{10000}{10000} = \frac{25}{10000} = \frac{1}{400}$; $.4 = .4 \times \frac{10}{10} = \frac{4}{10} = \frac{2}{5}$.

Ratios and Proportions

Proportions

A proportion is the link between one quantity and how it changes in relation to the changes in another quantity.

Direct proportion: a set increase in one quantity for every increase in the other quantity. Also, a direct proportion could be a set decrease in one quantity for every decrease in the other quantity. For example, if you increase the amount of water poured into a bottle, the weight of the bottle will increase incrementally in *proportion* to the amount of water being poured. If you decrease the volume of water in the bottle, the weight of the bottle will decrease.

Inverse proportion: a set increase in one quantity for every decrease in the other quantity. Also, an inverse proportion could be a set decrease in one quantity for every increase in the other quantity. For example, the darkness in a room decreases as light increases. Also, the amount of air in a cup increases as the water level decreases.

Ratios

Ratios quantify the comparison of two quantities. If there are 12 tacos and 6 people, the person to taco ratio is 6 to 12. This can be written as 6:12. Ratios should be reduced to their lowest whole number representation. In this case, 6:12 becomes 1:2 by dividing both sides by 6.

Probability

Probability is the likelihood of an event taking place. The higher the probability, the more likely the event is to take place. The lower the probability, the less likely the event is to take place. Probabilities are expressed as a number or fraction between 0 and 1.

To demonstrate this, consider a coin. When flipped, it can land on heads or tails. Since there are only two possible events, the probability of the coin landing on heads is 1 in 2 (.5, ½, 50%). Another classic example is a six-sided die roll. 6 possible events exist when rolling the die. The chance of rolling a 2 is 1 in 6 (⅙). Rolling any given number is a 1 in 6 chance. The probability of an event occurring is often calculated using the following equation:

$$P(A) = \frac{Number\ of\ acceptable\ outcomes}{Number\ of\ possible\ outcomes}$$

The total number of acceptable outcomes must be less than or equal to the total number of possible outcomes. If the total number of acceptable outcomes is equal to the total number of possible outcomes, then the probability of the event happening is certain and equal to 1. If the number of acceptable outcomes is zero, then the probability of the event occurring is impossible, and the probability is equal to 0.

Measures of Central Tendency

Data tends to cluster toward the middle of probability distributions. Central tendency describes these middle values. Three common measurements are taken to find these middle or "average" values. These common measurements are mean, median, and mode.

Range

When given a data set of numbers, the *range* is the difference between the highest and lowest number. For example, if the given data set is {2, 4, 9, 12, 19, 40}, the range is 40-2 = **38**.

Mean

The mean is what is commonly referred to as the average of a data set. Simply add up all the numbers in the data set and divide by the quantity of values in the data set.

$$Mean = \frac{Sum\ of\ data\ values}{Quantity\ of\ data\ values}$$

For example, if the data set is {4, 3, 9, 4, 1}, the formula to find the mean would be as follows:

$$\frac{4+3+9+4+1}{5} = \frac{21}{5} = 4.2$$

Mean doesn't always give an accurate picture of what the data is communicating. If you were given a data set of ages amongst 5 people that looked like this: {7, 5, 4, 9, 99}, the average age would technically be 24.8, which really does not depict the age of the group very well. The 99 year old person is what is known as an outlier, a value that is far outside the majority of the values. To communicate a more comprehensive picture of the data, the median and mode are used in conjunction with the mean.

Median

The median is simply the middle value of a data set. For example, if the data set is {17, 15, 20, 22, 14}, begin by rearranging the data from smallest to largest: {14, 15, 17, 20, 22}. Since the quantity of values is odd (five total values), then the middle value (17) is the median. If instead of five values there were six values, then the median would be the mean of the two middle numbers. For example, if the data set had a 16 in it ({14, 15, 16, 17, 20, 22}), the median would be $\frac{16+17}{2}$=16.5.

Mode

The mode is one more measurement of a data set. The mode is simply the value that appears the most number of times. If all values appear the same number of times, then there is no mode. If one value appears more times than any other value, then that value is the mode. If two or more values appear an equal number of times in respect to one another but more than the rest of the values, then those values are the modes.

Here are three examples:

1. The mode of this data set is 5 since it appears most frequently: {5, 4, 7, 4, 5, 6, 8, 5, 2, 3, 1, 2}

2. This data set has no mode because all values appear an equal number of times: {2, 4, 1, 6, 7, 9}

3. The mode of this data set is 3 and 5 because those two values appear an equal number of times in respect to one another, yet more than the rest of the data set: {3, 7, 5, 9, 5, 3, 8, 4}

Real World (Multiple Step) Problems

Example 1

A patient in the emergency room is given 1000 mL of fluids every two hours. What quantity of fluids will the patient receive after 5 hours?

Using proportional reasoning, since 5 hours is 2.5 times as long as 2 hours, then the patient will receive 2.5 times the amount of IV fluids, so 2.5 x 1000 = 2500 mL in 5 hours.

To compute this answer methodically, begin by writing the amount of IV fluids per 2 hours as a proportion:

$$\frac{1000 mL}{2 \text{ hours}}$$

Then, create a proportion to relate the two time increments, labeling the unknown quantity with x:

$$\frac{1000 \text{ } mL}{2 \text{ hours}} = \frac{x \text{ } mL}{5 \text{ hours}}$$

Make sure when you make a proportion to keep the numerator's and denominator's units consistent with each other (in this case, mL up top and hours in the bottom).

$$\frac{1000 \text{ } mL}{2 \text{ hours}} = \frac{x \text{ } mL}{5 \text{ hours}}$$
$$1000(5) = 2x$$
$$5,000 = 2x$$
$$5,000 \div 2 = 2x \div 2$$
$$2500 = x$$

So, this patient should receive 2500 mL of fluids every 5 hours.

Example 2

In one yoga class, there are 12 female students and 9 male students. The evening class, which is much larger, happens to have the exact same student ratio. If there are 48 female students in the evening class, how many male students are there?

$$\frac{\text{number of female students}}{\text{number of male students}} = \frac{12}{9} = \frac{48}{\text{number of male students}}$$

Then, write the number of male students as x.

$$\frac{12}{9} = \frac{48}{x}$$
$$12x = 48(9)$$

$$12x = 432$$
$$12x \div 12 = 432 \div 12$$
$$x = 36$$

Example 3

In a restaurant, there are 4 servers for 12 tables. If the restaurant were to expand to 24 tables, how many servers would they need to hire to ensure the quality of service does not suffer?

The ratio of servers to tables can be written as 4 to 12, or as 4:12, or as $\frac{4}{12}$. Since 12 and 4 share a common factor of 3, we can reduce this ratio to 1 to 3, or 1:3, or $\frac{1}{3}$. If this ratio were to remain constant, in order to cover $3 \times 8 = 24$ tables, they will need $1 \times 8 = 8$ servers. Since the restaurant already has 4 servers, they will need to hire 4 more.

Example 4

What is 20% of 160?

In this problem, the word *of* tells us that multiplication is happening, so you can find 20% of 160 by multiplying 160 by 20%. Convert the percentage into a fraction or decimal: $20\% = 0.20 = \frac{20}{100} = \frac{1}{5}$. Then, multiply:

$$160 \times 0.20 = 32,$$
$$\text{or:}$$
$$160 \times \frac{1}{5} = \frac{160}{5} = 32.$$

Example 5

What is 130% of 70?

In this problem, the word *of* tells us that multiplication is happening, so you can find 130% of 70 by multiplying 70 by 130%. Convert the percentage into a fraction or decimal: $130\% = 1.30 = \frac{130}{100} = \frac{13}{10}$. Then, multiply:

$$70 \times 1.30 = 91,$$
$$\text{or:}$$
$$70 \times \frac{13}{10} = \frac{910}{10} = 91.$$

Example 6

An employee is making $88 per day and then given a raise. After the raise, the employee is making $100 per day. What was the percentage increase of the employee's daily wages?

The wages increased by $12:
$$\$100 - \$88 = \$12$$

To find a percent increase (or decrease), we use $\frac{new\ amount\ -\ original\ amount}{original\ amount}$:
$$\frac{100-88}{88} = \frac{12}{88} = 0.136136 \approx 13.6\%$$

Basic Geometry

Geometry is the area of mathematics that deals with the shape, size, and relationships between objects in defined spaces. The most basic unit in geometry is the point.

A **point** is a fixed location. Since it is a place and not a thing, it has no size or dimension. It is represented by a dot and is usually labelled with an uppercase letter.

A **line** is defined by two points. A line is a figure that passes through both points, has no beginning and no ending, and extends in both directions.

A number of points which lie upon the same line are said to be **collinear.**
A **line segment** is a portion of a line that has two endpoints.

A **ray** has one endpoint and then continues infinitely in the other direction.

These are visually demonstrated on the next page:

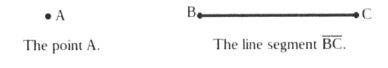

The point A.   The line segment $\overline{BC}$.

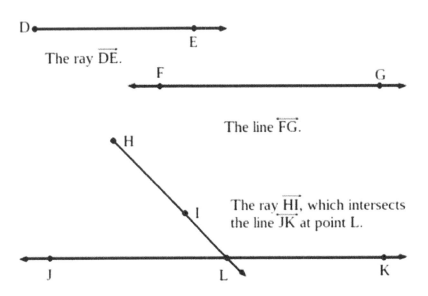

The ray $\overrightarrow{DE}$.

The line $\overleftrightarrow{FG}$.

The ray $\overrightarrow{HI}$, which intersects the line $\overleftrightarrow{JK}$ at point L.

A **plane** is a flat, two-dimensional surface extending infinitely in both directions. Just as a line may be defined by two points, a plane may be defined by three points.

A number of points, lines, rays, or line segments which lie on the same plane are said to be **coplanar**.

Parallel lines are defined as being two or coplanar lines which have no points in common and never meet.

Two lines which have exactly one point in common are called **intersecting lines**. More than two lines which have exactly one point in common are called **concurrent lines**.

A **transversal** is a line which intersects at least two other lines. Transversals which intersect parallel lines are useful tools in geometry, so we use them often.

When two lines or line segments intersect at a single point, they can be said to intersect at an **angle**. Angles are commonly represented by the symbol ∠. The point at which two lines, segments, or rays intersect to form an angle is called the **vertex** of that angle.

A **right angle** is an angle whose measure is 90 degrees.

Lines that intersect at an angle of 90 degrees (a right angle) are said to be **perpendicular**:

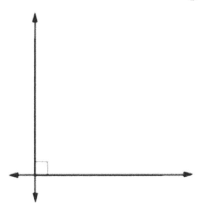

An **acute angle** is an angle that measures less than 90 degrees:

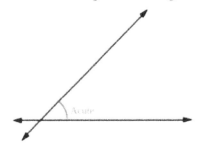

An **obtuse angle** is an angle that measures greater than 90 degrees, but less than 180 degrees:

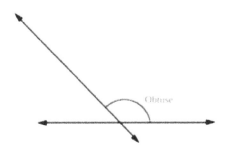

A **straight angle** is an angle that measures exactly 180 degrees. It looks like a straight line and sweeps out a semicircle, or half-circle.

A **reflex angle** is an angle that measures greater than 180 degrees, but less than 360 degrees:

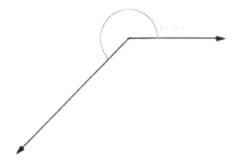

A **full angle** measures 360 degrees and sweeps out a full circle.

Classifying Figures

Squares and Rectangles

Squares and rectangles are **quadrilaterals**, or four-sided figures, made of perpendicular and parallel lines. If you take a square or rectangle and pick two sides going different directions, those lines are perpendicular. This tells you that all the angles in a square or rectangle are right angles! If you pick two lines going the same direction, those lines are parallel.

Parallelograms

Parallelograms have the word parallel in the name, so you shouldn't be surprised to find parallel lines in them:

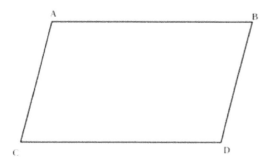

In the parallelogram shown in the figure, the sides AB and CD are parallel, and the sides AC and BD are parallel, too. In a parallelogram, each pair of opposite sides are parallel. Squares and rectangles are a special kind of parallelogram, but not all parallelograms are squares or rectangles.

A parallelogram whose sides are all the same length is called a **rhombus**.

Trapezoids

A trapezoid is a figure with two parallel sides and two sides that intersect those parallel lines. It has no perpendicular lines. In the given trapezoid, the lines AB and CD are parallel, but the lines AC and BD intersect the parallel lines:

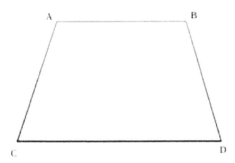

Triangles

A triangle is a figure with three sides. Triangles will never have any parallel lines. However, sometimes a triangle may contain perpendicular lines. Such a triangle is called a **right triangle**.

Right Triangles

A right triangle is a triangle in which two of the sides form a right angle:

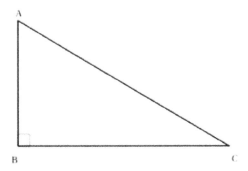

In a right triangle, the length of the longest side (AC) is called the **hypotenuse**. If you know the length of two of the three sides of a right triangle, you can figure out the length of the third side using the **Pythagorean Theorem**:

$$a^2 + b^2 = c^2$$

In this formula, a and b are the lengths of the shorter two sides, and c is the length of the hypotenuse.

Note: When talking about angles, the letter in the middle is the point where the angle happens. For instance, in the triangle figure, the angle ABC is a right angle, but the angle ACB is not - it's acute.

Areas of Polygons

Polygon: a figure composed of a shape with distinct sides and angles.

Vertex: in a polygon, any point where two sides intersect at an angle.

Altitude: the vertical height of a polygon.

Rhombus: a parallelogram whose sides are all congruent.

Kite: a quadrilateral whose pairs of adjacent sides are congruent.

Interior angle: the angle formed by the vertex of two sides of a polygon, measured in the interior of the polygon.

Regular polygon: a polygon whose sides and interior angles are congruent.

***n*-gon**: a polygon with *n* sides (for example, an 11-gon is a polygon with 11 sides).

Radius of a regular polygon: the distance from the center of a regular polygon to any of its vertices.

Central angle of a regular polygon: the angle formed by two radii leading to adjacent vertices.

Apothem: a line segment which bisects the central angle of a regular polygon, perpendicular to the side it intersects.

Squares and Rectangles

The area of a square is equal to the square of the length of one side, or simply $A = s^2$.

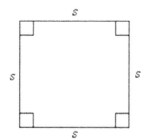

The area of a rectangle is similar to the area of a square; it is equal to the product of its length and width. You may also think of this as "base times height" or "side times side." Typically, the longer side is designated as the "length," and the shorter side is the "width," and so for a rectangle, you can write $A = lw$.

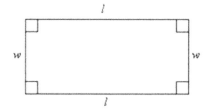

To find the area of a triangle, first choose one of the sides to be the base, b. Then, by taking a line perpendicular to the base rising up to the vertex above it, we find the altitude, or height, h, of the triangle.

The area of any triangle is equal to one half of the product of the base and the height, so your formula is $A = \frac{1}{2}bh$.

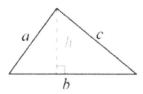

In order to determine the area of a parallelogram, you'll need to combine what you already know about rectangles and triangles. Notice that if you draw a perpendicular line from one corner of the parallelogram to its base, you can create a right triangle with altitude *h*. Doing the same from the other corner creates a rectangle with length *l* and width *h*.

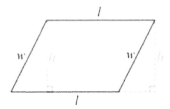

By "cutting off" the right triangle on the left and pasting it into the triangle-shaped area on the right, we will have a rectangle that must have the same area as the parallelogram. So, the area of the parallelogram is simply the area of this rectangle: $A = lh$.

Area of a Rhombus or Kite

A rhombus is a special kind of parallelogram whose sides are all the same length. In order to find the area of a rhombus, you need to find the lengths of its diagonals. In the given figure, the rhombus has diagonals *a* and *b*.

Notice that these diagonals divide the rhombus into four right triangles, each with base ½ *b*, and height ½ *a*. Based on what you know about triangles, each of these triangles will have an area of *(1/8) ab*. Since the combined areas of these four congruent triangles is the same as the area of the rhombus, the area of the rhombus must be $4(\frac{1}{8}ab) = \frac{1}{2}ab$. So, for a rhombus, $A = \frac{1}{2}ab$.

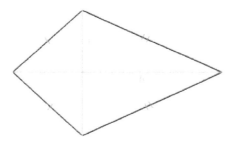

A kite is a quadrilateral with two pairs of congruent adjacent sides. A kite differs from a rhombus in that it is not a parallelogram, but because it can be broken up into four right triangles with the use of its diagonals, the area of a kite can be found using exactly the same formula as the rhombus. So, for a kite, $A = \frac{1}{2}ab$.

Let's review the areas in a single table:

Square	$A = s^2$
Rectangle	$A = lw$ (where l is length and w is width)
Triangle	$A = \frac{1}{2} bh$ (where b is the base, and h the altitude or height)
Parallelogram	$A = lh$ (where l is the length, and h is the altitude or height)
Rhombus	$A = \frac{1}{2} ab$ (where a and b are the lengths of its diagonals)
Kite	$A = \frac{1}{2} ab$ (where a and b are the lengths of its diagonals)

Angle Measure in Regular Polygons

A polygon whose angles and side lengths are all equal is called a *regular polygon*. A polygon that isn't regular is called *irregular*. This is a regular hexagon (left), and an irregular hexagon (right):

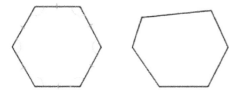

Polygons are named according to the number of sides they have. Since the polygon in the figure has six sides, it's a hexagon. The hexagon on the left has all its sides congruent, and all its interior angles are congruent. Every angle in this regular hexagon will have the same measure. However, if you were to draw a regular octagon (that is, a regular eight-sided polygon), the interior angles, while congruent with each other inside the octagon, would be different from those in the hexagon.

In order to learn how to find the measure of the interior angles of a regular polygon, consider this table. All polygons in this table are considered to be regular polygons.

Number of Sides	Regular Polygon Name	Each Interior Angle Measure
3	Triangle	60°
4	Quadrilateral	90°
5	Pentagon	108°
6	Hexagon	120°
7	Heptagon	128.57°
8	Octagon	135°
9	Nonagon	140°
10	Decagon	144°
...	...	...
n	n-gon	$\dfrac{(n-2)180°}{n}$

Areas of Regular Polygons

Just as a circle has a radius, a regular polygon will also have a radius. The distance from the central point to any of the vertices is the *radius* of a regular polygon. The angle formed by two radii to two adjacent vertices in a regular polygon is called the *central angle*.

Notice in the figure the segment labeled *a*. This is a line segment which bisects the central angle and is perpendicular to the side it intersects. This is called the *apothem* of a regular polygon.

Taking repeated radii and apothems splits the regular *n*-gon into a number of triangles, so you can use similar methods to the one used for the rhombus to determine that the area of a regular *n*-gon is one half the product of the apothem's length and the polygon's perimeter.

So, for a regular *n*-gon, $A = \frac{1}{2} pa$.

Three-Dimensional Figures

We encounter solids every day in our real lives. A can of soup is basically a cylinder, and a brick is basically a cuboid (that is, a six-sided solid with rectangular sides). By studying geometric solids, we can learn about the physical world around us.

In geometry, a solid is a three-dimensional figure that has width, height, and depth. Some classic examples include the cube, sphere, cylinder, cone, and pyramid.

Here are some examples of geometric solids:

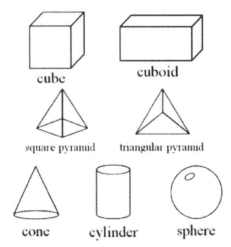

- **Cube**: a solid whose faces are square. It has six faces, eight vertices, and 12 edges.

- **Cuboid**: a solid whose faces are rectangular. It has six faces, eight vertices, and 12 edges.

- **Square Pyramid**: a solid with a square-shaped face on the bottom, and four triangular faces adjacent. It has five faces, eight edges, and five vertices.

- **Triangular Pyramid**: a solid whose faces are triangles. It has four faces, six edges, and four vertices.

- **Cone**: Starting with a circular base, a cone is formed by all line segments that join a point on the circle to a single locus, called the apex of the cone. A cone has two faces, one vertex, and one edge.

- **Cylinder**: A cylinder has circular bases at the top and bottom, and a curved surface in between. You can think of this as being like a single circle "stretched" along a third dimension, or a pair of circles joined by all line segments connecting corresponding points on the circles.

- **Sphere**: defined by a central point P, and every point that is equidistant (at a radius of R) from the point across three dimensions. It looks like a ball or marble.

Surface Area of a Prism

You know by now how to find the area of various polygons. How does this apply to solids? Given a solid, you can determine its surface area – that is, the combined area of all its faces – and its volume, the quantity of 3d space enclosed by the solid.

To start with, let's consider the surface area of a *prism*. A prism is a particular kind of solid that has a polygon base on one "end," a second base which is merely a translated copy of the first, and n faces joining corresponding sides of the two bases.

A cuboid is a kind of prism. In order to find the surface area of a prism, you need to determine two things: the base area, and the lateral area.

The base area, denoted B, is the area of one of the bases.

The lateral area, denoted L, is the combined area of all lateral faces – that is, faces that join the two bases. In our cuboid example, all the lateral faces (and the two bases) are rectangles.

Then, you can determine the surface area S by using the following formula:

$S = 2B + L.$

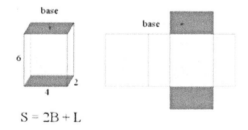

In general, the volume of a prism can be expressed as $V = Bh$, where B is the area of the base, and *h* is the height.

Cylinders

Since the bases of a cylinder are both circles, you only need to know its radius to determine their area. Since there are two bases, the area of the "ends" of the cylinder is $2(\pi r^2)$.

For any piece of the cylinder between the two ends, the length around is $2\pi r$ (the circumference of a circle), and so the surface area of the exterior face is $A = 2\pi rh$ because you can unwrap this surface and create a rectangle with length $2\pi r$ and width of

h. Adding the area of the outside face to the area of the two bases gives the surface area of the whole cylinder. So, the surface area of a cylinder is given by $S = 2(\pi r^2) + 2\pi rh$.

Finding the volume of a cylinder follows a similar procedure:

$$V = \pi r^2 h$$

Volume of a Pyramid

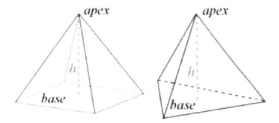

Unlike the comparatively complex prism, pyramids obey a fairly simple rule when it comes to finding their volume. The volume of a pyramid can be found using a single formula:

Volume of a Pyramid = (1/3) * Area of Base * height

$$V = \left(\frac{1}{3}\right) Bh$$

Cones

A cone has a feature called the *axis*, which is an imaginary line that passes through the center of the circular base and is perpendicular to the base plane. If you imagine the base of the cone as a wheel, the axis would be its axle.

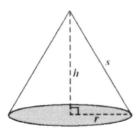

The surface area of a right cone will always be given by: $S = \pi r^2 + \pi rs$

A cone looks sort of like a pyramid, but with circles and sectors instead of polygons and triangular faces. That comparison is valid, it turns out, because the formula for the volume of a cone is exactly the same as the formula for the volume of a pyramid:

$$V = \left(\frac{1}{3}\right) Bh$$

Since the base of a cone is always a circle, then $B = \pi r^2$, and $V = \left(\frac{1}{3}\right) \pi r^2 h$.

Spheres

On a sphere, a **great circle** is a circle whose center includes the point at the center of the sphere. On Earth (modeled as a perfect sphere), the Equator and the Prime Meridian are both great circles.

If you can determine the area of a great circle in a sphere, take that area and multiply it by 4 to get the surface area of the entire sphere:

$$S = 4\pi R^2.$$

The volume of a sphere is given by $V = (4/3)\pi R^3$.

Extras

Common Denominator and Least Common Multiple

The least common multiple of two or more numbers is the smallest number which is a multiple of all of those numbers. For example, multiples of 4 include 4, 8, 12, 16, 20, 24, 28, and 32. The multiples of 6 include 6, 12, 18, 24, 30, and 36. So, the least common multiple of 4 and 6 is 24.

When two fractions are changed in order to have the same denominator, this is called finding a **common denominator**. The common denominator will usually be the **least common multiple** of the two starting denominators. For example, consider $\frac{4}{5}$ and $\frac{6}{8}$. The least common multiple of 5 and 8 is 40, so to reach a common denominator, $\frac{4}{5} = \frac{32}{40}$ and $\frac{6}{8} = \frac{30}{40}$.

Greatest Common Factor

A factor is a number that will exactly divide some other number. For example, the factors of 12 are 1, 2, 3, 4, 6, and 12. The factors of 13 are 1 and 13 - its only factors are 1 and itself. Any number larger than 1 with this characteristic is called a **prime number**. Some common prime numbers are 2, 3, 5, 7, 11, 13, and 17. Note that 2 is the only even prime number.

A **common factor** is a number that exactly divides two or more other numbers. The factors of 15 are 1, 3, 5, and 15. The factors of 9 are 1, 3, and 9. The common factors of 15 and 9 are 1 and 3.

A **prime factor** is a factor which is also prime. The prime factors of 15 are 3 and 5. The **greatest common factor** of two or more numbers is the greatest number that is a factor of all of those numbers. For example, the factors of 24 are 1, 2, 3, 4, 6, 8, 12, and 24. The factors of 36 are 1, 2, 3, 6, 12, 18, and 36. The greatest common factor of 24 and 36 is 12.

Ordering of Rational Numbers

Recall that a rational number is a number which can be expressed as a ratio or fraction $\frac{a}{b}$ where a and b are integers.

Example: order the following rational numbers from least to greatest: $0.6, 23\%, \frac{28}{4}, \frac{5}{8}, 2$.

We can write each of these numbers as a fraction, or as a decimal:

$$0.6 = 0.6$$
$$23\% = 0.23$$
$$\frac{28}{4} = 7$$
$$\frac{5}{8} = 0.625$$
$$2 = 2.$$

Reordering them, we get:

$$23\% = 0.23$$
$$0.6 = 0.6$$
$$\frac{5}{8} = 0.625$$
$$2 = 2.$$
$$\frac{28}{4} = 7$$

So, the correct ordering is $23\%, 0.6, \frac{5}{8}, 2, \frac{28}{4}$.

Before You Start the Practice Questions

On the next page, you will begin an Arithmetic Reasoning practice test. Set a timer for 29 minutes before you start this practice test. Giving yourself 29 minutes will give you an authentic feel for how long you have to finish this portion of the AFOQT. Like the official test, this practice test has exactly 25 items. Set your timer, turn the page, and begin.

Arithmetic Reasoning Practice Test Questions

(29 Minutes)

1. A car travels 250 miles in 5 hours. How far can the car travel in an hour and a half?

    a. 25 miles
    b. 50 miles
    c. 75 miles
    d. 100 miles
    e. 125 miles

2. June buys $60 worth of apples and bananas. Apples cost $1 each, and bananas cost $2 each. If she buys the same number of apples and bananas, how many bananas did she buy?

    a. 1
    b. 5
    c. 10
    d. 15
    e. 20

3. Find the surface area of a box with square sides 5 feet long, 5 feet wide, and 5 feet tall.

    a. 50 square feet
    b. 75 square feet
    c. 100 square feet
    d. 125 square feet
    e. 150 square feet

4. Find the volume of a box with rectangular sides 4 feet wide, 3 feet long, and 2 feet tall.

    a. 12 cubic feet
    b. 16 cubic feet
    c. 24 cubic feet
    d. 30 cubic feet
    e. 36 cubic feet

5. A ship travels 40 nautical miles per hour (40 knots) and sails in a straight line for 2 hours. A sea chart has a scale of 1 inch per 10 nautical miles. How many inches apart are the ship's starting and ending points on the chart?

   a. 2 inches
   b. 4 inches
   c. 6 inches
   d. 8 inches
   e. 10 inches

6. A school has 12 teachers and 18 teaching assistants. The school has 150 students. What is the ratio of faculty to students?

   a. 1:5
   b. 12:18
   c. 18:150
   d. 15:70
   e. 1:4

7. Samantha has two part-time jobs. At one, she earns $21,000 per year. At the second, she earns $12,000 per year. After putting $100 each month into savings, Samantha donates 10% of her remaining income to charity. How much does she donate every year?

   a. $3000
   b. $3180
   c. $3250
   d. $3425
   e. $3300

8. A box with rectangular sides measures 18 inches wide, 12 inches deep, and 6 inches high. Find the volume of the box in cubic feet.

   a. .85 cubic feet
   b. .75 cubic feet
   c. 2 cubic feet
   d. 1.5 cubic foot
   e. .5 cubic feet

9. John weighs 180 pounds and wants to lose weight. His goal weight is 150 pounds. About how many kilograms does he need to lose? (1 pound is approximately 0.45 kilograms.)

   a. 9 kg
   b. 10 kg
   c. 13.5 kg
   d. 16 kg
   e. 20 kg

10. The perimeter of a regular hexagon is 42 cm. What is the length of one side?

    a. 5 cm
    b. 7 cm
    c. 9 cm
    d. 11 cm
    e. 13 cm

11. Five students take a test. Four of them score 80, 85, 80, and 90. If the average score is 82, what did the fifth student score?

    a. 75
    b. 80
    c. 85
    d. 90
    e. 70

12. In a carnival game of skill involving tossing a ball into a scoring bucket, Chandra makes 15 attempts and scores 8 times. What percentage of Chandra's balls went into the bucket?

    a. 49%
    b. 53%
    c. 61%
    d. 65%
    e. 71%

13. Randy knows that he has ⅞ of an ounce of food for his fish. His friend gives him another bag of fish food, and a scale tells Randy that the new amount is 0.125 ounces. How much fish food, in ounces, does Randy have total? You may convert from fraction to decimal or from decimal to fraction to solve.

   a. 1 ounce
   b. 2 ounces
   c. 3 ounces
   d. 4 ounces
   e. 5 ounces

14. An amusement park adds an automated trolley to take visitors from one area to another. The trolley travels 75 yards in 15 seconds. How long (in seconds) will it take the trolley to carry a guest from the Spook House to the Screaming Falcon roller coaster if they are 225 yards from each other and the trolley does not need to stop in between?

   a. 35 seconds
   b. 40 seconds
   c. 45 seconds
   d. 50 seconds
   e. 55 seconds

15. Palermo is a pitcher and is working on his fastball. Using a radar machine to track the speed of his pitches, he throws six fastballs that register at 87, 89, 93, 91, 80, and 85 miles per hour. Based on this practice session, what is the mean speed of Palermo's fastball?

   a. 85 mph
   b. 86 mph
   c. 87.5 mph
   d. 88 mph
   e. 89.5 mph

16. Siobhan grows tomatoes in her backyard vegetable garden. Last year, her single tomato plant produced 8 pounds of tomatoes. This year, she tried a different soil, and her single tomato plant produced 11 pounds of tomatoes. By what percent did her tomato production increase?

   a. 30%
   b. 33%
   c. 35%
   d. 37.5%
   e. 39%

17. Josh wants to carpet his home office. His home office consists of a large room, which is 12 feet by 10 feet, and a smaller storage room, which is 8 feet by 8 feet. How much carpet (in square feet) will he need to carpet both rooms in his home office?

    a. 150 square feet
    b. 175 square feet
    c. 180 square feet
    d. 184 square feet
    e. 190 square feet

18. June marks out a triangle-shaped area for an art mural. Before she paints the mural, she needs to cover the area with primer. The triangle's base is 12 feet long, and it is 5 feet tall. How much primer (in square feet) will she need to cover the triangular area?

    a. 20 square feet of primer
    b. 25 square feet of primer
    c. 30 square feet of primer
    d. 35 square feet of primer
    e. 40 square feet of primer

19. In a parcel of land, Robert fences off a large rectangular field and splits that field into 3 smaller fields. If the large field is 30 feet by 200 feet, what length of fence does Robert need to fence off the outside and the dividing lines inside? (Hint: draw a picture of this!)

    a. 400 feet of fence
    b. 470 feet of fence
    c. 480 feet of fence
    d. 500 feet of fence
    e. 520 feet of fence

20. Tulsi has a solid box-shaped container with an open top which is 3 feet long, 4 feet wide, and 4 feet tall. What is the surface area of her container?

    a. 50 square feet
    b. 55 square feet
    c. 60 square feet
    d. 68 square feet
    e. 73 square feet

21. Now suppose Tulsi wished to fill her cuboid container with water. What volume of water (in cubic feet) would she need?

   a. 48 cubic feet
   b. 50 cubic feet
   c. 52 cubic feet
   d. 55 cubic feet
   e. 60 cubic feet

22. Lin has a cylindrical shaped water tank in his home lab. It has a radius of 12 inches and a height of 36 inches, with a lid sealing the top. If Lin wanted to paint the tank and the lid, how much paint (in square inches) would he need?

   a. 3,222 square inches
   b. 3,619 square inches
   c. 4,155 square inches
   d. 4,777 square inches
   e. 5,312 square inches

23. Now suppose that after he had painted his cylindrical tank, Lin wished to fill it with water. How much water (in cubic inches) would it hold?

   a. 16,286 cubic inches
   b. 22,555 cubic inches
   c. 24,670 cubic inches
   d. 27,345 cubic inches
   e. 32,345 cubic inches

24. Parvati designs a cone-shaped coffee filter for a commercial coffee machine. The filter fits a receptacle that is shaped like a cone with a radius of 4 inches and a depth of 4 inches. How much coffee, in cubic inches, could she theoretically fit inside her coffee filter?

   a. 54 cubic inches
   b. 67 cubic inches
   c. 55 cubic inches
   d. 65 cubic inches
   e. 57 cubic inches

25. In a neighborhood, the median household income is $40,000 per year. The highest paid person in the neighborhood gets a promotion which comes with a $15,000 per year raise. What is the value for the new median household income?

   a. $55,000
   b. $50,000
   c. $45,000
   d. $40,000
   e. $35,000

Answer Guide to Arithmetic Reasoning Practice Test

1. **C**. A car travels 250 miles in 5 hours, so its speed is $250 \div 5 = 50$ miles per hour. In 1.5 hours, the car will travel $50 \times 1.5 = 75$ miles.

2. **E**. Let x be the number of apples and bananas she bought. Then, $60 = 1x + 2x$.
$$60 = 3x$$
$$20 = x$$
So, she buys 20 bananas (and 20 apples).

3. **E**. The box has six sides. Each side has an area of $5 \times 5 = 25$ square feet. $25 \times 6 = 150$ square feet in total surface area.

4. **C**. The volume is $l \times w \times h = 4 \times 3 \times 2 = 24$ cubic feet.

5. **D**. The ship travels a total of $40 \times 2 = 80$ nautical miles. On the map, this corresponds to a distance of $80 \div 10 = 8$ inches.

6. **A**. There are $12 + 18 = 30$ total faculty. The ratio of faculty to students is 30:150, or 1:5.

7. **B**. At $100 per month, each year Samantha puts $\$100 \times 12 = \$1200$ into savings. Her total annual income is $\$21,000 + \$12,000 = \$33,000$. Subtracting what she puts into savings, Samantha has $\$33000 - \$1200 = \$31800$ left. If she donates 10% of this to charity, then she donates $\$31800 \times 0.10 = \$3180$ to charity each year.

8. **B**. First we need to convert the dimensions from inches to feet. Since 1 foot is 12 inches, we can divide each dimension by 12 to get its length in feet.
$$18 \div 12 = 1.5$$
$$12 \div 12 = 1$$
$$6 \div 12 = 0.5$$
Then, the volume of the box is $l \times w \times h = 1.5 \times 1 \times 0.5 = 0.75$ cubic feet.

9. **C**. John wants to lose $180 - 150 = 30$ pounds. This is approximately $30 \times .45 = 13.5$kg.

10. **B**. A regular hexagon has six sides of equal length, so each side measures $42 \div 6 = 7$cm.

11. **A**. If the average is 82, then the sum of all of their scores is $82 \times 5 = 410$. To find the missing score, subtract the others.
$$410 - 80 - 85 - 80 - 90 = 75.$$
The fifth student scored 75.

12. **B.** $\frac{8}{15} = 53\%$

13. **A.** We will convert to decimal: ⅞ = 0.875. Then, 0.875 + 0.125 = 1 ounce.
If we instead convert to fractions, 0.125 = ⅛, and so ⅛ + ⅞ = 1 ounce.

14. **C.** Let's split the journey into 15-second increments. In the 225 yard path there are
225 ÷ 75 = 3 15-second increments for a travel time of 15 × 3 = 45 seconds.

15. **C.** 87 + 89 + 93 + 91 + 80 + 85 = 525 ÷ 6 = 87.5mph.

16. **D.** Her total increase is 11-8 = 3 pounds of tomatoes. Percent increase is
$\frac{increase}{original\ amount} = \frac{3}{8} = 0.375$ or 37.5%.

17. **D.** He will need 12 × 10 = 120 square feet in the large room and 8 × 8 = 64 square feet in the smaller room, for a total of 120 + 64 = 184 square feet.

18. **C.** The area of a triangle is $\frac{1}{2}bh = \frac{1}{2}(12)(5) = \frac{1}{2}(60) = 30$ square feet of primer.

19. **E.** He needs the perimeter of the field, plus the length of the dividers.

The large field is 30 by 200, and each of the 2 dividers must also be 30 feet. So, he needs a total of 2(200) + 4(30) = 400 + 120 = 520 feet of fence.

20. **D.** The box has a bottom which is 3 feet by 4 feet, two sides which are 3 feet by 4 feet, and two sides which are 4 feet by 4 feet. It has no top. Find the area of the bottom and four sides, and add them together to find surface area.
(3)(4) + (3)(4) + (3)(4) + (4)(4) + (4)(4) = 12 + 12 + 12 + 16 + 16 = 68 square feet.

21. **A.** The fact that the box does not have a top does not impact the volume of water it will hold. This is also $V = l \times w \times h = 3 \times 4 \times 4 = 48$ cubic feet.

22. **B**. The surface area of a cylinder is $S = 2(\pi r^2) + 2\pi rh = 2\pi(12)^2 + 2\pi(12)(36) \approx$ 3,619 square inches.

23. **A**. The volume of a cylinder is $V = \pi r^2 h = \pi(12)^2(36) \approx 16{,}286$ cubic inches.

24. **B**. The volume of a cone is $V = \left(\frac{1}{3}\right) Bh = \frac{1}{3}(\pi r^2 h) = \frac{1}{3}\pi(4)^2(4) \approx 67$ cubic inches.

25. **D**. This is sort of a trick question. The median is not affected by an increase of the maximum data point, even if that change would change the range and mean. So, the median household income is still $40,000 per year.

WORD KNOWLEDGE

This section of the AFOQT will measure your vocabulary knowledge. In the Verbal Analogy section, your communication skills were tested by analyzing your ability to find relationships and contrasts between different words. This section is quite different. You will have to rely on your vocabulary knowledge. You will be given a word and you must choose the synonym or word that most closely defines the given word. Before you start the practice test, we will go over some methods of eliminating wrong answers in cases in which you do not know the definition of the given word.

The Structure of the Question

The structure of word knowledge questions is quite simple. You are given one word and then a list of possible answers. Choose the word that means the same thing or the word that has the closest definition to the given word. This is very different from the verbal analogies section. In this section, you are only looking for words with similar meanings. This is an example of a word knowledge question:

Humid
   a. Wet
   b. Humble
   c. Steamy
   d. Juicy
   e. Exhausting

Every question on the word knowledge section will appear like the one above. You are given one word (in this case, "Humid"), then you look through the options and choose the word with the most similar meaning. In the above example, the answer is "C. Steamy." "Humid" is an adjective that describes a large amount of water vapor in the atmosphere. "Steamy" is synonymous with that meaning. Answer choice "A. Wet" may be tempting to choose, but it is not as close in meaning.

How to Prepare for This Section

The most challenging part about this section is that it is unpredictable. The Air Force does not provide a list of words that you can study. You are forced to rely on the vocabulary and word knowledge you have acquired before the test. One helpful step in preparing would be to go online and search for flash cards or word lists for standardized tests, such as the SAT. So many books and online resources are available to expand your overall vocabulary for standardized tests. It is impossible to know which words you will see on the AFOQT, but broadening your vocabulary with online lists and flashcards will give you better odds of succeeding on this section of the test.

Remain confident even if you doubt your ability to recall word meanings. The reality is that you've been speaking for most of your life, and you have a general understanding of how words are formed and their potential meanings. Even if you do not recognize the word, you may be able to break it down and make an inference about the meaning of the word. We will cover a handful of strategies to approach questions in instances where you do not readily know the given word's meaning.

Prefixes

Pay attention to the prefix of the given word. Prefixes make up the first part of a word. If you encounter a word you do not know, looking at the prefix may give you a clue as to what the word means. Look at the example below:

Uncommon
    a. Weird
    b. Popular
    c. Shared
    d. Communication
    e. Shake

The word "uncommon" has the prefix *un-*. The prefix *un-* means *not*. If something is uncommon, it is literally "not" common. From choices A through E, you would look for a word that means "not common." "Weird" fits that definition better than any other word above. The answer is indeed "A. Weird."

Some prefixes can have multiple meanings or share a similar meaning with another prefix, so don't completely rely on your understanding of prefixes to understand a word's definition. For instance, the prefix *sub-* and *under-* can mean the same thing (think "submarine" or "underneath"), but the prefix *sub-* can also mean "nearly" (as in "subtropical").

Sometimes an answer will share a similar prefix as the given word. Look at the next question and choose the correct answer:

Misinterpret
   a. Analyze
   b. Misunderstand
   c. Wander
   d. Flounder
   e. Exhaust

"B. Misunderstand" is the correct answer. "Misinterpret" means to wrongly interpret. "Misunderstand" means to wrongly understand. The two words have a similar meaning. Once again, pay attention to prefixes to give you a hint as to what the right choice will be. Having the same prefix does not make a word the right answer, but be aware of the prefix to help you make the right choice.

You may already know the meaning of many prefixes, but below you will find a list of common ones to review. This is only a short list of frequently used prefixes, but knowing these is a good start:

Prefix	Meaning	Example(s)
a-, an-	without, not, lacking	atypical, anarchy
ambi-	both	ambivalent
ante-	before, earlier, in front of	antecedent
com-, con-	together, with	combine, consensus
contro-, contra-	against, opposite	controversy, contradict
dis-	apart, away	disquiet
en-	within, cover with	enclose
ex-	out of, from	exhale
extra-	beyond, outside	extracurricular
il-, im-, in-, ir-	not, without	illicit, immoral, incomplete, irregular
inter-	between, among	interchange
omni-	all, every	omnipotent
post-	after, later	posthumous
pre-	before	precedent
sym-, syn-	together	symmetrical, synthesis
trans-	across, through	transaction, translucent

Origin and Roots of Words

The English language is primarily derived from Greek and Latin words. Sometimes you can decipher the correct answer by recognizing root words. Consider the following example:

Insomnia
- a. Anxiety
- b. Sleeplessness
- c. Energetic
- d. Awake
- e. Fast

The Latin words *in* and *somnis* together form the Latin word *insomnis,* which means *sleepless.* "Insomnia" is an English word that means "sleeplessness," so your best answer choice is B. The rest of the options may be characteristics of insomnia, but only sleeplessness is synonymous.

Knowing the meaning of Latin words is helpful, but sometimes just recognizing the root words without knowing the meaning can help. Examine this question:

Acrimonious
- a. Wonderful
- b. Acrid
- c. Fearless
- d. Precarious
- e. Truthful

"Acrimonious" and "acrid" both have the same Latin root word (*acri-*). *Acri* in Latin means *sharp or pungent.* "Acrimonious" and "acrid" can both mean "angry and bitter," although "acrid" is often used to describe a bad smell. Compared to the options above, "acrid" has the most similar definition to "acrimonious." Even if you don't know the definition of "acrimonious," choosing an answer that shares a Latin root word may be the correct answer, as is the case in this sample question.

If you would like to build your knowledge of root words, many good resources are available online. For example, you can do a search for "root words" on sites like *yourdictionary.com* or find root word lists on *vocabulary.com.*

Similarities

Sometimes the relationship between answers can give you a hint as to what the correct answer is. This is one technique that should only be used if you do not know the meaning of the given word. This strategy involves eliminating words that have similar meanings to one another. Consider the following:

Insularity
    a. Tired
    b. Exhausted
    c. Ignorance
    d. Sleepy
    e. Worn

"Insularity" is most closely defined as "ignorance." If you have no clue as to what "insularity" means, you may be able to guess that it means "ignorance" if the rest of the word definitions are so similar. Be careful with this strategy, as it has no guarantee of being foolproof. It is more of a last resort strategy. Tired, exhausted, sleepy, and worn have similar meanings, but "ignorance" is completely different. You may be able to guess that "C. Ignorance" is right since it is unlikely that four words could be so closely tied in meaning to the given word. More importantly, if "insularity" did mean "tired," it may also mean sleepy/worn/exhausted, thus decreasing the odds that the testmaker would make one of two similar words the correct answer. Since "ignorance" does not share the same meaning as the other options, it is logically more likely to be the correct answer. Also, "insularity" and "ignorance" are the only nouns (more on this in the next strategy). Keep in mind that this is just another tool for your toolbox.

Sentence Technique

Placing the given word in a sentence and then replacing it with each word option can shed light on the correct answer. Observe the example below:

Flawless
    a. Good
    b. Perfect
    c. Best
    d. Incredible
    e. Fantastic

Form a sentence with the word flawless: "Her skin is flawless." No word options fit as well as "B. Perfect." Even though some fit pretty well, only the word "perfect" is synonymous in meaning.

Does the Word Sound Positive or Negative?

Often, words are positive sounding or negative sounding. Words such as clumsy, dishonest, and evil all have negative connotations. On the flipside, words such as beautiful, charming, and fantastic are positive. This is a quick way to decipher a correlation between the given word and the possible word choices. Examine this question:

Knowledgeable
   a. Ghastly
   b. Haggard
   c. Insidious
   d. Intelligent
   e. Repugnant

"Knowledgeable" is a fairly positive sounding word with a positive connotation. Most of us think highly of knowledge. Additionally, the affix *-able* gives a positive feeling (in contrast to the more negative state of being *un*able). Ghastly, haggard, insidious, and repugnant all sound negative. It may seem silly, but they simply sound negative when you say them out loud. Even if you do not know their meanings, you may get a sense for whether or not they have a positive connotation. None of them have a similar meaning, but you could still infer the correct answer, "D. Intelligent," without knowing their definitions because "Intelligent" is the only positive sounding word. Obviously, "intelligent" and "knowledgeable" share a similar definition, but they also share a positive connotation. Pay attention to the connotation of words and whether or not they sound negative or positive. You may be surprised by how many words in the English language sound positive or negative. Of course, not all words sound like their connotations, so use this strategy only when you don't know the meanings of the words.

Analyzing Parts of Speech

Pay attention to the parts of speech when answering the questions in this section of the test. Is the given word a noun, verb, adjective, or adverb? Usually, the correct answer will be the same part of speech as the given word. You may have learned about these basic parts of speech in school, but here are some simple guidelines for identifying them easily:

Nouns: You probably remember a noun as a "person, place, or thing." Anything that can be counted is a noun (for example, I have one computer. "Computer" is a noun). However, some nouns are uncountable, like "happiness." You can't count it, but it is considered a thing (an abstract thing).

Fortunately, there are some *suffixes* (word endings) that are most common to nouns. Below is an incomplete list of common noun suffixes, along with some examples. When you see words that end in these suffixes, you are usually looking at nouns:

- *-ness* (happiness, friendliness, resourcefulness)
- *-ion* (dimension, fascination, hesitation)
- *-ity* (hilarity, enmity, velocity)
- *-age* (wreckage, adage, baggage)
- *-dom* (freedom, kingdom, stardom)
- *-ship* (friendship, hardship, sportsmanship)

Verbs: Verbs are called "action words" because they usually describe the action that someone or something is doing. They are also the only words that have tense (past, present, future, etc.). If you can change the tense of the word in a sentence, it is a verb. For example, imagine you are given the word "infuriate." You can create different sentences using that word in different tenses: they infuriate me, they infuriated me, they will infuriate me, etc. You will be given words out of context (so not in sentences) on this section of the test, but you can make up simple sentences in your mind to test for verbs. Even if you don't know the word's meaning, you will be surprised at how often you will still be able to tell when a word makes grammatical sense in a sentence.

Adjectives: Adjectives are "describing words." Specifically, they describe nouns. If it grammatically fits in the frame "The _____ person," or "The _____ thing," it is an adjective. Like nouns, some suffixes are common to adjectives. When you see words that end in these suffixes, you are often looking at adjectives:

- *-able/-ible* (knowledgeable, forgettable, convertible)
- *-ous* (porous, gorgeous, momentous)
- *-ful* (forgetful, bountiful, beautiful)
- *-less* (guileless, spineless, heartless)
- *-ive* (festive, effective, talkative)
- *-al* (national, hormonal, logical)

Adverbs: Adverbs are describing words too, but they describe verbs, adjectives, and other adverbs. For example, in the sentence, "She runs quickly," "quickly" is an adverb describing the verb "runs." In the sentence, "He is frequently late," "frequently" is an adverb describing the adjective "late."

The most common suffix of an adverb is -*ly*. You saw it in the examples above and can see it in many other adverbs, such as smoothly, harshly, candidly, etc.

Knowing words' parts of speech can help you when you are eliminating answer choices (we will discuss how soon). Remember, however, that these words on the test will be without context, so they can still be tricky. For example, a word like "insult" can be a noun or verb, depending on how it's used. In the sentence "That's quite an insult," it's a noun. In the sentence "They insulted me," it's a verb. However, this knowledge will still give you an edge when you're looking for synonyms.

Now let's practice. Consider the following:

Sadness
    a. Downcast
    b. Unhappiness
    c. Droopy
    d. Gloomy
    e. Woeful

If you are given a question like this on the test, pay attention to the parts of speech. All the options above describe sadness, but only "B. Unhappiness" shares the same part of speech. "Sadness" is a noun. "Unhappiness" is a noun. The rest are all adjectives. In this case, "unhappiness" would be the correct answer. If "unhappiness" was not synonymous with "sadness," then it would not be the right answer. The part of speech only matters when you are given options that have nearly identical meanings.

Good vs. Best

For each question, you will be given a word and a list of options to choose from. Some of the options may be good choices, but only one is the best selection. Consider the following example:

Joyful
    a. Happy
    b. Excited
    c. Content
    d. Elated
    e. Animated

Most of the choices above are good options, but "A. Happy" is the *best* option. "Happy" has the closet definition to "joyful." "Elated" is an extreme form of happiness and joy.

"Content" is one form of joy. Much like "elated," "excited" is a stronger form of "joyful" and "happy." "Happy" is synonymous with "joyful," even though the rest of the options may appear to be correct.

Word Recall

Sometimes you might see a word and have no idea what it means. In those moments, the word may remind you of one that you already know. For instance, if you did not know what the word "mythical" or "mythology" meant, but you knew what the word "myth" meant, you could infer that "mythical" or "mythology" relates to something fictitious. You may deduce the meaning of the root word by thinking of another word you know with a similar word structure. Obviously, just because one word reminds you of another does not mean it has a similar meaning. This is only a strategy to use if you have absolutely no idea what option to select. Often, one word may remind you of a word that has no correlation; nonetheless, this is another tool to put in your toolbox if you have no idea which option to select.

Final Thoughts

All these strategies are only to be used if you do not know the meaning of the given word or options. Ideally, you will know the definition to the given word and be able to choose the word that is most similar in meaning. Only use these techniques if they are needed.

Before You Start the Practice Questions

On the next page, you will begin a Word Knowledge practice test. Set a timer for 5 minutes before you start this practice test. Giving yourself 5 minutes will give you an authentic feel for how long you have to finish this portion of the AFOQT. Like the official test, this practice test has exactly 25 items. Set your timer, turn the page, and begin.

Word Knowledge Practice Test Questions

(5 Minutes)

1. Ambivalent
    a. Working
    b. Predominate
    c. Undecided
    d. Incipient
    e. Dexterity

2. Benediction
    a. Contrite
    b. Blessing
    c. Veneer
    d. Barren
    e. Tonic

3. Vehement
    a. Passionate
    b. Pious
    c. Succor
    d. Torment
    e. Hoax

4. Sage
    a. Cognate
    b. Suffice
    c. Secular
    d. Wise
    e. Burnish

5. Blithe
    a. Stingy
    b. Hallow
    c. Presumptuous
    d. Repine
    e. Thoughtless

6. Cryptic
    a. Mysterious
    b. Squeaky
    c. Itinerate
    d. Acarpous
    e. Churl

7. Incoherent
    a. Transient
    b. Edible
    c. Confusing
    d. Boring
    e. Squat

8. Prodigal
    a. Sermon
    b. Wasteful
    c. Recast
    d. Sash
    e. Pristine

9. Scanty
    a. Pariah
    b. Hateful
    c. Sponge
    d. Minimal
    e. Exorbitant

10. Verbose
    a. Nonchalant
    b. Slow
    c. Buoyant
    d. Active
    e. Wordy

11. Tact
    a. Tape
    b. Intrigue
    c. Diplomacy
    d. Suppress
    e. Antidote

12. Reprisal
    a. Credulous
    b. Odor
    c. Grief
    d. Retaliation
    e. Stray

13. Prowess
    a. Skill
    b. Sting
    c. Eradicate
    d. Installation
    e. Prey

14. Intrigue
    a. Disprove
    b. Succor
    c. Simper
    d. Fascinate
    e. Restive

15. Fervor
    a. Deposition
    b. Enthusiasm
    c. Coerce
    d. Insularity
    e. Truce

16. Depravity
    a. Corruption
    b. Pliant
    c. Frugal
    d. Coddle
    e. Implosion

17. Brevity
    a. Refractory
    b. Garner
    c. Sublime
    d. Conciseness
    e. Droll

18. Qualm
    a. Brittle
    b. Peace
    c. Unease
    d. Wave
    e. Reproach

19. Rapport
    a. Affinity
    b. Catalyst
    c. Rumple
    d. Pique
    e. Undulate

20. Sect
    a. Hew
    b. Group
    c. Revolt
    d. Shovel
    e. Stickler

21. Probity
    a. Importance
    b. Polished
    c. Chisel
    d. Fulsome
    e. Integrity

22. Respite
    a. Lurk
    b. Rest
    c. Violent
    d. Supercilious
    e. Hurry

23. Solace
    a. Stubborn
    b. Talkative
    c. Grueling
    d. Comfort
    e. Sun

24. Volition
    a. Sanity
    b. Will
    c. Holster
    d. Electricity
    e. Prudence

25. Amass
    a. Wean
    b. Turn
    c. Accumulate
    d. Inept
    e. Delight

Answer Guide to Word Knowledge Practice Test

1. **C.** "Undecided" has the closest meaning to "ambivalent" from the options provided.
2. **B.** "Blessing" is synonymous with "benediction." *Bene* in Latin means *well. Dicere* in Latin means *say. Benedicere* in Latin would mean *wish well/bless.*
3. **A.** "Passionate" is synonymous with "vehement." "Torment" has the same suffix as "vehement," but its definition has more to do with severe suffering. "Passionate" is the best choice.
4. **D.** "Wise" is synonymous with "sage." No other options have a similar meaning.
5. **E.** "Thoughtless" is synonymous with "blithe." No other options are even close to the same meaning.
6. **A.** "Mysterious" is synonymous with "cryptic." *Crypticus* comes from the Greek word *kruptikos*, meaning *hidden.*
7. **C.** "Confusing" is synonymous with "incoherent." No other options have a similar meaning.
8. **B.** "Wasteful" is synonymous with "prodigal." *Prodigus* in Latin means *lavish.* The word "prodigal" describes someone who spends money in a lavish and reckless manner.
9. **D.** "Minimal" is synonymous with "scanty." Although "exorbitant" and "scanty" are often used with a negative connotation, "exorbitant" is quite the opposite from the words "minimal" or "scanty." It means "excessive" rather than "meager."
10. **E.** "Wordy" is synonymous with "verbose." *Verbum* in Latin means *word.* Someone who is verbose may be "active" in how they speak, but active is not synonymous with "verbose."
11. **C.** "Diplomacy" is synonymous with "tact." No other options have a similar meaning.
12. **D.** "Retaliation" is synonymous with "reprisal." No other options have a similar meaning.
13. **A.** From the available options, "skill" and "prowess" have the most similar meaning.
14. **D.** "Fascinate" is synonymous with "intrigue."
15. **B.** "Enthusiasm" is synonymous with "fervor."
16. **A.** "Corruption" is synonymous with "depravity." *Pravus* in Latin means *crooked or perverse*.
17. **D.** "Conciseness" is synonymous with "brevity." *Brevis* in Latin means *brief.*
18. **C.** "Unease" is synonymous with "qualm." No other options have a similar meaning.
19. **A.** "Affinity" and "rapport" are very close in meaning. Both "affinity" and "rapport" can describe a relationship between people.

20. **B.** "Group" has the closet definition to "sect" from the listed options. "Sect" comes from the Latin words *sequi* and *secta*. *Sequi* means *follow* and *secta* means *following*. "Revolt" is something that a sect might do, but the two words do not mean the same thing.
21. **E.** "Integrity" is synonymous with "probity." *Probus* in Latin means *good*.
22. **B.** "Rest" is synonymous with "respite." *Respectus* in Latin means *refuge*.
23. **D.** "Comfort" is synonymous with "solace." *Solari* in Latin means *to console*.
24. **B.** "Will" is synonymous with "volition." No other options have a similar meaning.
25. **C.** "Accumulate" is synonymous with "amass."

MATH KNOWLEDGE

The Math Knowledge portion of the AFOQT is more challenging than the Arithmetic Reasoning section. It may test your knowledge of factors, polynomials, exponents, square roots, and other concepts. Study for the Math Knowledge section of the AFOQT much like you did for the Arithmetic Reasoning section: read through this chapter, take the practice test, and then compare your answers with those in the answer guide.

Exponents

In math, we use exponents as a way to write products of repeated factors.

For example, if we write $2 \cdot 2 \cdot 2 \cdot 2 \cdot 2 \cdot 2$, we are essentially taking the factor "2" and multiplying it together 6 times. This can also be written as 2^6, where 2 is called the "base" and 6 is called the "exponent." The term 2^6 is called an exponential expression, or a *power*.

So in general terms, $a^n = a \cdot a \cdot a \cdot a \cdot ... \cdot a, where\ there\ are\ n\ factors\ of\ a.$

If we wanted to read or say an exponential expression using words, we can do this in a couple of different ways. Going back to our example of 2^6, we would read or say that as "two to the sixth power" or "two raised to the power of six."

Two very common exponents are 2 and 3. When we use the exponent 2, we refer to this as "squared," and when we use the exponent 3, this is called "cubed."

When we are working with exponents in either an expression or an equation, there are some specific rules that must be followed in order to solve them correctly. We need to consider multiplication properties of exponents, division properties of exponents, zero exponents, and negative exponents. Let's take a look at each type:

Multiplication Properties:

- Product of Powers Property: When you want to multiply powers with the same base, add the exponents.

    For example, $a^5 \cdot a^7 = a^{5+7} = a^{12}$

- Power of a Power Property: To find the power of a power, multiply the exponents.

    For example, $(a^2)^4 = a^{2 \cdot 4} = a^8$

- Power of a Product Property: To find the power of a product, find the power of each factor separately, and then multiply them together.

    For example, $(a \cdot b)^3 = a^3 \cdot b^3$

Division Properties:

- Quotient of Powers Property: When you want to divide powers with the same base, subtract the exponents.

    For example, $\frac{a^{12}}{a^8} = a^{12-8} = a^4$

- Power of a Quotient Property: To find the power of a quotient, find the power of the numerator and denominator separately and then divide them.

    For example, $\left(\frac{a}{b}\right)^7 = \frac{a^7}{b^7}$

Zero Powers: If a is a nonzero number, then $a^0 = 1$.

   For example, $6^0 = 1 \; and \; 127^0 = 1$

Negative Powers: If a is a nonzero number, then $a^{-n} = \frac{1}{a^n}$.

   In other words, a^{-n} is the reciprocal of a^n.

   For example, $a^{-4} = \frac{1}{a^4}$

Also, it should be noted that any number raised to the power of 1 is equal to that number.

   Example: $12^1 = 12$

Bases can be negative or positive. For example, $2^3 = 8$ has a positive base of 2. However, let's say the base was -2. This would give us $(-2)^3 = -8$. There is a simple rule of thumb to remember when you have a negative base, which, in this case, is -2. If the power is even (2,4,6...), then the result will be positive. If the power is odd (3, 5, 7...), then the result will be negative.

   For example, $(-3)^2 = 9$ and $(-3)^3 = -27$.

Exponents are commonly used in science when working with either very large or very small numbers. In order to express these numbers more easily, we use something called *scientific notation.*

Scientific notation is expressed as $a \times 10^n$, where $1 \leq a < 10$ and n is an integer. In other words, a is being multiplied by a power of 10.

To write a number in scientific notation, place a new decimal point to the right of the first nonzero digit. From this new decimal point, count the number of digits to the original decimal point location. This number will represent the exponent, n. If the original number is a very small number (less than zero), then you will be counting digits to the left and that will make n a negative number. Alternatively, if the original number is a very large number, then you will be counting digits to the right, making n positive.

See the following examples:

$$1{,}580{,}000 = 1.58 \times 10^6$$

$$10000 = 1.0 \times 10^4$$

$$0.000047 = 4.7 \times 10^{-5}$$

Roots

In the previous section we talked about exponents, and one of the exponents we talked about was 2, or "squaring" a number. The inverse, or opposite, operation to squaring a number is taking its square root.

For example, $5^2 = 25$. So the square root of 25 would be equal to 5.

We write a square root with the $\sqrt{}$ symbol. This is called a *radical symbol,* and the number under the radical is called the *radicand.* So in the above example, we would write the square root of 25 as $\sqrt{25}$, which equals 5. However, it should be noted that if we squared (-5), that would also equal 25, so $\sqrt{25} = 5 \; and -5$ or ± 5.

Therefore, all positive real numbers have two roots: a positive square root and a negative square root.

There are two types of square roots: *perfect squares* and *imperfect squares.*

Perfect squares are numbers that, when you take the square root, the result is an integer. 0, 1, 4, 9, 16, 25, 36, 49, 64, 81, and 100 are all examples of perfect squares.

For example, $0^2 = 0$; $1^2 = 1$; $2^2 = 4$; $3^2 = 9$; $4^2 = 16$, and so forth.

So $\sqrt{0} = 0$; $\sqrt{1} = 1$; $\sqrt{4} = 2$; $\sqrt{9} = 3$; $and \sqrt{16} = 4$.

Imperfect squares cannot be easily solved and require a calculator. See the following examples:

$$\sqrt{14} = 3.74$$

$$\sqrt{45} = 6.71$$

We can estimate imperfect squares by looking at which perfect squares are closest to the imperfect square on either side. So in the example above, $\sqrt{45}$ is between the perfect squares for $\sqrt{36} = 6$ and $\sqrt{49} = 7$. Therefore, we know that the answer to $\sqrt{45}$ is between 6 and 7.

While square roots are the most common root we see, there are others you should be familiar with, such as the cube root, which is written as $\sqrt[3]{}$.

An example of a cube root is $\sqrt[3]{27} = 3$, since $3^3 = 27$.

Parentheses

Parentheses are used in expressions and equations to group certain terms together. The parentheses require that the expression or equation be solved in a certain order. We will talk about order of operations for solving an expression or equation in the next section. For now, we will just talk about an expression or equation that is strictly comprised of parentheses.

There are different types of parentheses, sometimes called brackets. They are $(\ )$, $[\ ]$, and $\{\ \}$.

When solving an expression or equation with multiple parentheses, the rule is to start with the innermost parentheses and work your way out to the outermost parentheses.

Let's look at an example:

$\{10 - [(5 + 2) \cdot 1]\}$ Start with the innermost parentheses $(5 + 2)$.

$\{10 - [7 \cdot 1]\}$ Now move on to the next innermost parentheses $[7 \cdot 1]$.

$\{10 - 7\}$ Now finish with the outermost parentheses to get an answer of 3.

Order of Operations

All expressions and equations must be solved using a set of rules known as the *order of operations.*

The rules are as follows:

Step 1: Solve anything within parentheses first.

Step 2: Then solve anything with exponents second.

Step 3: Next, solve any multiplication and/or division from left to right.

Step 4: Finally, solve any addition and/or subtraction from left to right.

This is commonly referred to as PEMDAS: Parentheses, Exponents, Multiplication, Division, Addition, and Subtraction. Some people will come up with a phrase or sentence that will help them to remember this order. A common one is "Please Excuse My Dear Aunt Sally," where the first letter of each word represents the first letter of each of the operations in order.

It is important to note that during multiplication/division and addition/subtraction, another rule applies called the *Left to Right Rule.* This rule states that when multiplication and division or addition and subtraction exist by themselves, you should solve from left to right.

Let's look at a few examples:

Example 1:

$2 \cdot 3^2 + 5$	There are no parentheses in this expression, so move on to the exponents.
$2 \cdot 9 + 5$	Now solve the multiplication.
$18 + 5$	The answer is 23.

Example 2:

$10 - 3 + (2 + 5)$	First, solve what's inside the parentheses.
$10 - 3 + 7$	Now we have only addition and subtraction, so the left to right rule takes priority. Solve the subtraction next.
$7 + 7$	The answer is 14.

Example 3:

$6 \div 3 + 2 \cdot 7$	Since there are no parentheses or exponents, solve the multiplication first.
$6 \div 3 + 14$	Next, solve the division.
$2 + 14$	The answer is 16.

Real Numbers

All real numbers are either positive, negative, or zero.

You can visualize the real numbers on a number line like this:

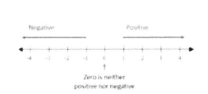

Points to the left of zero are negative. These numbers are marked with a (-) symbol to show that they are negative.

Points to the right of zero are positive. Usually, these numbers are not marked with any symbol.

As you can see, zero is neither positive nor negative.

Practice with the number line in the following example:

Name each of the points on the number line above.

On this number line, we can see that between each number are 3 lines that create 4 sections. What this means is that each line is $\frac{1}{4}$, or 0.25. Notice that as we move to the right, the numbers get closer to 0 because we are on the negative side of the number line. Let's identify each point now:

A: This point is halfway between -5 and -6, which makes it -5.5.

B: The dot is exactly on -4.

C: The dot is on the line directly to the right of -3. So it is at -2.75.

D: This point is 1 line to the left of -1, which makes it -1.25.

We can also use the number line to learn about something called *absolute value*.

Absolute value of a number is defined as its distance away from zero on the number line. It is shown as the number enclosed by ||.

Let's look at a couple of examples:

$|-2| = 2$ We can see that the absolute value of -2 is equal to 2, because -2 is two units away from zero on the number line.

$|4| = 4$ The absolute value of 4 is equal to 4, because 4 is four units away from zero on the number line.

Let's look at one more example:

$-|1| = -1$ If you have a negative sign that occurs outside of the absolute value bars, you should apply the negative sign after determining the absolute value of the enclosed number. So in this example, 1 is one unit away from zero, so the absolute value is equal to 1. But we have a negative sign outside of the absolute value bars, so we apply that negative sign to our value of 1. Therefore, the final answer is -1.

Operations with Real Numbers

Addition:

If the two numbers have the same sign, then add the numbers and keep the sign.

For example:

$3 + 8 = 11$ Two positive numbers added = positive number.

$-2 + (-7) = -9$ Two negative numbers added = negative number.

Please note that if you are adding two negative numbers, you actually take the absolute value of each number, add them together, and then apply the negative sign to the answer. Let's look at the last example again:

$$-2 + (-7) = -9$$
$$|-2| + |-7|$$
$$2 + 7 = 9$$

Now apply the negative sign to the answer. The final answer is -9.

When you are adding two numbers with opposite signs, you subtract the numbers and keep the sign of the bigger number. Please note: Always take the absolute value of the numbers and then subtract them.

Let's look at a couple of examples:

$$-3 + 5 = ?$$

$|-3| = 3 \text{ and } |5| = 5$ First, take the absolute values of the numbers.

$5 - 3 = 2$ Then subtract those numbers.

Finally, assign the sign of the bigger number of the original numbers, 3 and 5, to the final answer.

The final answer is 2. Because the bigger number is 5, and it was positive, we assign a positive sign to the final answer.

Let's try another one:

$$4 + (-7) = ?$$

$|4| = 4 \text{ and } |-7| = 7$ Take the absolute values of the numbers.

$7 - 4 = 3$ Subtract those numbers.

Assign the sign of the bigger of the original numbers, 4 and 7, to the answer. Since 7 is the bigger number, and it was originally negative, then the final answer will be negative. So the final answer is -3.

Subtraction:

When subtracting two numbers, we don't actually subtract; rather, we do what's called "adding the opposite." We will use something called *Keep – Change – Change:*

Keep the sign of the first number before the subtraction.

Change the subtraction symbol to addition.

Change the sign of the second number to the opposite sign.

Let's look at an example:

$5 - (-12) = ?$

$5 + 12 = ?$ Using Keep – Change – Change, we keep the sign of the 5, change the subtraction to addition, and change the (-12) to 12.

The answer is 17.

Let's look at one more example:

$-3 - 6 = ?$

$-3 + (-6) = ?$ Using Keep – Change – Change, we keep the sign of the (-3), change the subtraction to addition, and change the 6 to (-6).

The answer is -9.   Notice that we now have two negative numbers that we are adding together. We can use our rules for addition now to solve this.

Multiplication & Division:

When multiplying positive and negative numbers the rules are as follows:

$$(+) \cdot (+) = (+)$$
$$(+) \cdot (-) = (-)$$
$$(-) \cdot (-) = (+)$$

The rules for division follow the rules for multiplication.

If you have more than two numbers that you are multiplying/dividing, then count how many negative numbers you have, and if it is an odd number, the answer will be negative, and if it is an even number, then the answer will be positive.

Let's look at a few examples:

$-4 \cdot 8 = -32$ $(-) \cdot (+) = (-)$

$-15 \div (-3) = 5$ $(-) \div (-) = (+)$

$10 \cdot (-2) \cdot 6 \cdot 1 = -120$ Since there are four numbers being multiplied together, we can see that there is only one occurrence of a negative number. Therefore, the number of negative numbers is odd, so the answer is negative.

In this section on operations with real numbers, we have so far only looked at operations with integers. However, operations with real numbers can also include fractions and decimals. Let's look at some examples:

Example 1:

Simplify $\frac{4}{5} + \frac{3}{4}$

In this problem we need to find a common denominator between the two fractions so that we can add them. To find a common denominator, we write out the multiples of each denominator and look for the least common multiple. This will become our common denominator.

Multiples of 5:   5, 10, 15, 20, 25, 30, ...

Multiples of 4:   4, 8, 12, 16, 20, 24, 28, ...

You can see that the first multiple that is common between the two denominators is 20. This is the least common multiple, or *LCM,* and will be our least common denominator, or *LCD.*

Next, we need to convert each of our fractions into new equivalent fractions with the new common denominator:

$\frac{4}{5} = \frac{16}{20}$: To get 5 to 20, we need to multiply by 4. Whatever we do to the denominator, we want to do the same to the numerator. This creates a new, equivalent fraction. Therefore, the numerator becomes $4 \cdot 4 = 16$.

$\frac{3}{4} = \frac{15}{20}$: To get this new equivalent fraction, we need to multiply both the numerator and denominator of the original fraction by 5.

Now that we have the new equivalent fractions with a common denominator, we can add them:

$$\frac{16}{20} + \frac{15}{20} = \frac{31}{20}$$

Example 2:

Simplify $\frac{3}{8} - \frac{1}{6}$

This subtraction problem with fractions can be solved using steps similar to the addition problem in example 1.

Find a common denominator first:

Between the denominators 8 and 6, the least common multiple is 24. This is our new common denominator.

Find new equivalent fractions using the new common denominator:

$\frac{3}{8} = \frac{9}{24}$ and $\frac{1}{6} = \frac{4}{24}$

Now we can subtract the two fractions:

$\frac{9}{24} - \frac{4}{24} = \frac{5}{24}$

Example 3:

Simplify $\frac{\frac{3}{7}}{\frac{5}{14}}$

This is a division of fractions problem. It could also have been written as: $\frac{3}{7} \div \frac{5}{14}$.

When you have a division of fractions problem, you complete a couple of steps to solve it:

First, change the division into multiplication.

Next, flip the second fraction to its reciprocal.

Finally, solve the new multiplication problem by multiplying across and reducing the fraction if possible.

So to solve this example as a multiplication problem, the new expression becomes:

$$\frac{3}{7} \cdot \frac{14}{5}$$

Now, multiply across and reduce if possible: $\frac{3}{7} \cdot \frac{14}{5} = \frac{42}{35} = \frac{6}{5}$

Example 4:

Simplify $\frac{5}{6} + \left(2\left(\frac{3}{5}\right) - 1\right) + 2$

To solve this problem, we need to use the order of operations. We should solve within the innermost parentheses first and work out from there.

The innermost operation is $2\left(\frac{3}{5}\right)$.

When you have a whole number multiplied by a fraction, you can rewrite the whole number a as $\frac{a}{1}$, which means the same thing because any number divided by 1 is the original number.

$$\frac{2}{1} \cdot \frac{3}{5} = \frac{6}{5}$$

So now we have $\frac{5}{6} + \left(\frac{6}{5} - 1\right) + 2$.

To solve what's within the next set of parentheses, we need to change the 1 to an equivalent fraction with a denominator of 5: $1 = \frac{5}{5}$.

Therefore, the expression becomes $\frac{5}{6} + \left(\frac{6}{5} - \frac{5}{5}\right) + 2 = \frac{5}{6} + \frac{1}{5} + 2$.

Now we are left with an addition problem. We need to find a common denominator to add these three terms. Remember, we can rewrite 2 as $\frac{2}{1}$.

The least common multiple between the terms is 30. This is our new common denominator.

Let's find equivalent fractions with the new common denominator, so that we can add them:

$$\frac{5}{6} = \frac{25}{30} \qquad \frac{1}{5} = \frac{6}{30} \qquad 2 = \frac{2}{1} = \frac{60}{30}$$

Note that when you are looking for an equivalent fraction of a whole number, the numerator will be bigger than the denominator. Finally, let's add the new equivalent terms:

$$\frac{25}{30} + \frac{6}{30} + \frac{60}{30} = \frac{91}{30}$$

Since there are no common factors between 91 and 30, we cannot reduce this fraction any further.

Example 5:

Simplify $0.18 + 1.43 - (0.61 + 3.27 \cdot 0.29)$

To solve this problem, we need to use the order of operations by solving within the parentheses first and working out.

Within the parentheses, we have addition and multiplication. According to the rules for the order of operations, we need to complete the multiplication first and then the addition.

So $(0.61 + 3.27 \cdot 0.29) = (0.61 + 0.9483) = 1.5583$,

The expression is now $0.18 + 1.43 - 1.5583$.

The expression contains addition and subtraction, and should be solved from left to right.

$$0.18 + 1.43 - 1.5583 = 1.61 - 1.5583 = 0.0517$$

Example 6:

Simplify $\frac{9}{4} + (2(0.32) - 0.15) - 1$

To solve this problem, we need to use the order of operations starting within the parentheses and working outward.

Within the parentheses is multiplication and subtraction. So we will solve the multiplication first and then complete the subtraction:

$$(2(0.32) - 0.15) = 0.64 - 0.15 = 0.49$$

The expression has become $\frac{9}{4} + 0.49 - 1$

We have addition and subtraction left. Solve the expression from left to right. It is easiest to write the fraction $\frac{9}{4}$ as a decimal in this problem: $\frac{9}{4} = 2\frac{1}{4} = 2.25$

$$2.25 + 0.49 - 1 = 2.74 - 1 = 1.74$$

Example 7:

Simplify $4.8 - 1.2 + (12 - 6.9 \div 3) + 16$

As we have done in all of our examples, we must use the order of operations to solve this problem.

First, we need to solve what is inside the parentheses: $(12 - 6.9 \div 3)$. Since we have subtraction and division, we do the division first because that is first in the order of operations. Then we will do the subtraction part. This becomes $(12 - 6.9 \div 3) = (12 - 2.3) = 9.7$.

The expression is now $4.8 - 1.2 + 9.7 + 16$.

Since we have only subtraction and addition, we will solve from left to right now:

$$4.8 - 1.2 + 9.7 + 16 = 3.6 + 9.7 + 16 = 13.3 + 16 = 29.3$$

Polynomials

A polynomial is a type of algebraic expression comprised of either one or a sum of terms. Each of these terms can be referred to as a *monomial*. Within a polynomial, there can only be whole number exponents and no variables in denominators.

An example of a polynomial is $3x^5 - 4x^3 + 8x + 10$.

This polynomial is comprised of the following terms or monomials: $3x^5, -4x^3, 8x,$ and 10.

Let's look at the leading term, $3x^5$. The 3 in this term is called the *coefficient*. It is called the *leading term* because it is the term with the highest exponent. It is customary to write a polynomial in the order of terms with the largest exponent first, followed by smaller and smaller exponents. The leading term can also tell us the degree of the polynomial. Since it has the largest exponent, which is 5, we know that this polynomial is a fifth-degree polynomial.

We can add and subtract polynomials by combining *like terms*. Like terms are terms that have the same variable with the same exponent. Let's look at an example:

$(3a^5 - 9a^3 + 4a^2) + (-8a^5 + 8a^3 + 2)$

We can combine $3a^5$ and $-8a^5$ because they have the same variable, a, and the same exponent, 5. When you combine like terms, you keep the variable and the exponent the same, and you add up the coefficients to get your new coefficient.

So. $3a^5 + (-8a^5) = -5a^5$.

Then we can combine $-9a^3$ and $8a^3$ because they share a common variable and exponent.

So. $-9a^3 + 8a^3 = -1a^3$.

We can't combine any other terms, so the answer is $-5a^5 + (-1a^3) + 4a^2 + 2$.

If we wanted to subtract polynomials, we would use the same process; just remember to distribute the subtraction symbol through to each term in the second polynomial.

For example, $(3r + 8) - (2r - 5)$ becomes $3r + 8 - 2r - (-5)$ or $3r + 8 - 2r + 5 = r + 13$.

Note that if the coefficient is 1, we can drop the 1 because it is implied. When you are adding terms together and you see a variable without a coefficient, you can assume the coefficient is 1. That way, you can combine it with like terms, if there are any. So $1r = r$.

We can also multiply polynomials together. To do this, we multiply each term in the first polynomial by each term in the second polynomial. We can use a method known as *FOIL* to help us do that.

Let's look at the following example:

$$(x^2 + 5x)(3x^4 + 2)$$

FOIL stands for First, Outer, Inner, Last. This means we multiply the first terms of the polynomials, then we multiply the outer terms of the polynomials, then we multiply the inner terms of the polynomials, and finally, we multiply the last terms of the polynomials. Let's try it for our example:

First: $x^2 \cdot 3x^4 = 3x^6$

Outer: $x^2 \cdot 2 = 2x^2$

Inner: $5x \cdot 3x^4 = 15x^5$

Last: $5x \cdot 2 = 10x$

Once we do this, we then write the remaining terms as a sum. This would be:

$$3x^6 + 2x^2 + 15x^5 + 10x$$

We then check to see if there are any like terms to combine. In this example, there aren't. Then the last step is to write the polynomial with the exponents in descending order.

The final answer is $3x^6 + 15x^5 + 2x^2 + 10x$.

We can also factor polynomials. To factor a polynomial is to write it as the product of two or more simpler polynomials.

The easiest way to factor is to find the GCF, or *greatest common factor,* between the coefficients and between the variables, and then multiply those results.

Let's look at the following example:

$$9x^4 + 15x^2 + 3x$$

First, we'll look at the coefficients 9, 15, and 3. Let's write out the *factors* of each number:

Factors of 9 = 1, 3, 9

Factors of 15 = 1, 3, 5, 15

Factors of 3 = 1, 3

Factors are a list of all of the numbers that you can multiply together to get the current number. So for 9, we can do $1 \cdot 9$ and $3 \cdot 3$, which means that the factors of 9 are 1, 3, and 9 because there are no other numbers we could multiply to get 9.

To find the GCF, take a look at the numbers and find the number or numbers that are shared between them. Then the GCF is simply the largest of those shared numbers. In this case, we only have one number that is shared: 3. So 3 is our GCF for the coefficients.

Next, we look at the variables. The GCF between $x^4, x^2,$ and x is simply x.

This is because we can rewrite these exponential terms like this:

$$x^4 = x \cdot x \cdot x \cdot x$$

$$x^2 = x \cdot x$$

$$x^1 = x$$

We can see that the greatest common factor between these numbers is x because that is what they share. Remember, if a variable has no exponent, that means its exponent is 1.

So now we take our GCF from our coefficients and our GCF from our exponential terms and multiply them together. This gives us a complete GCF of $3x$ between all three terms of the polynomial.

So, if we factor that out, the polynomial becomes $3x(3x^3 + 5x + 1)$.

The idea behind factoring is that you can re-multiply the GCF by each of the terms inside the parentheses and get the original term. For example, $3x \cdot 3x^3 = 9x^4$, which is the original term in the polynomial. When you factor out the GCF, try to think about what you need to have inside the parentheses in order to arrive at the original polynomial when you multiply everything out.

Here are some rules for factoring special types of polynomials that are very helpful to know:

Perfect Square Trinomials:

$$x^2 + 2xy + y^2 = (x + y)^2$$
$$x^2 - 2xy + y^2 = (x - y)^2$$

Difference of Squares:

$$x^2 - y^2 = (x + y)(x - y)$$

Difference of Cubes:

$$x^3 - y^3 = (x - y)(x^2 + xy + y^2)$$

Sum of Cubes:

$$x^3 + y^3 = (x + y)(x^2 - xy + y^2)$$

Perfect Cube Trinomials:

$$x^3 + 3x^2y + 3xy^2 + y^3 = (x + y)^3$$
$$x^3 - 3x^2y + 3xy^2 - y^3 = (x - y)^3$$

Rational Expressions

A rational expression is the quotient of two polynomial functions.

We can view this as: $\frac{a(x)}{b(x)}$, where $b(x) \neq 0$.

Some examples of rational expressions include:

$\frac{6x}{9x^2}$ $\frac{x^2+8x+16}{3x+12}$ $\frac{2x^2}{x(x+5)}$

We can simplify a rational expression in a similar way to simplifying a fraction. Simplify both the numerator and denominator by factoring and then canceling out any common factors. A rational expression is in its most simplified form when it doesn't contain any common factors between the numerator and denominator other than 1.

Let's look at an example:

$\frac{5x}{10x^2-5x}$ The first step is to factor the numerator, if possible. However, there is nothing to factor.

$\frac{5x}{5x(2x-1)}$ The second step is to factor the denominator. There is a common factor of 5x between the terms that we can factor out.

$\frac{1}{1(2x-1)}$ The third step is to cancel out any common factors between the numerator and denominator. We can see that 5x is common between both the numerator and denominator, so we can cross that out from both.

$\frac{1}{(2x-1)}$ This leaves the final answer with the rational expression in its simplest form.

We can also perform arithmetic operations on rational expressions. Addition, subtraction, multiplication, and division are done using the same rules for performing these operations on simple fractions.

Let's look at addition and subtraction first:

Remember that in order to add or subtract fractions, the fractions must have a common denominator. The same rule applies for adding or subtracting rational expressions.

If the denominators of the rational expressions are already the same, we simply add the numerators and keep the common denominator.

See the following example:

$\frac{4}{x+2} - \frac{x+4}{x+2}$ Since the denominators are the same, we can just add the numerators and keep the denominator.

$\frac{4-x-4}{x+2}$

Remember when you are subtracting to apply the subtraction symbol to all terms in the second expression.

$\frac{-x}{x+2}$

Next, simplify the numerator by combining any like terms. In this example, we can combine the 4 and -4. This leaves -x in the numerator.

The final step is to simplify the rational expression. In this example, the expression is in its simplest form. So this is the final answer.

Now let's look at a case where we want to add or subtract two rational expressions that don't have a common denominator.

As is the case with fractions, we need to find the least common denominator (LCD) of the two denominators in the rational expression.

To find the LCD:

Factor the denominators. Then take the highest power of each factor that appears in either denominator and multiply those together to get the LCD.

Let's look at an example:

$\frac{1}{36x} + \frac{3x+1}{9x^5}$ First, factor the two denominators, $36x$ and $9x^5$.

$$36x: 4 \cdot 9 \cdot x = 2^2 \cdot 3^2 \cdot x$$

$$9x^5: 9 \cdot x^5 = 3^2 \cdot x^5$$

If we take the highest power of each factor that appears in either denominator and multiply them together, we get: $2^2 \cdot 3^2 \cdot x^5 = 36x^5$.   This is our LCD.

Now we have to make each rational expression in our original addition problem have the LCD of $36x^5$.

We again use the same rules we would use to change any fraction to a LCD. We multiply both the numerator and denominator of each expression to obtain the LCD:

$\frac{1}{36x} \cdot \frac{x^4}{x^4} = \frac{x^4}{36x^5}$ and $\frac{3x+1}{9x^5} \cdot \frac{4}{4} = \frac{4(3x+1)}{36x^5} = \frac{12x+4}{36x^5}$

Now we can add the two rational expressions:

$\frac{x^4}{36x^5} + \frac{12x+4}{36x^5} = \frac{x^4+12x+4}{36x^5}$

We cannot simplify this rational expression any further, so the final answer is $\frac{x^4+12x+4}{36x^5}$.

Multiplication and division of rational functions also follow the same rules for multiplying and dividing fractions. Let's look at multiplication first.

When multiplying two fractions we multiply the numerators, multiply the denominators, and then simplify the fraction. We can apply the same rules to multiplying rational expressions. The only difference is that we first factor each rational expression to see if there are any terms that can be cancelled out prior to multiplying.

Let's look at an example:

$\frac{5x+10}{x-3} \cdot \frac{x^2-9}{5}$ First, we will factor the two rational expressions.

$\frac{5(x+2)}{x-3} \cdot \frac{(x-3)(x+3)}{5}$ Now that we have factored both rational expressions, we can see that there is a common 5 in both the numerator and denominator of the fractions, so that can be cancelled out. There is also a common (x-3) in both the numerator and denominator that can be cancelled out.

$\frac{(x+2)}{1} \cdot \frac{(x+3)}{1}$ This leaves us with these two rational expressions to be multiplied.

$\frac{(x+2)(x+3)}{1}$ After multiplying across, the final answer is $(x+2)(x+3)$.

Division of rational expressions is done in a similar way. But first, we need to change the division into a multiplication problem. Remember that when you are dividing two fractions, you flip the second fraction to its reciprocal and change the division into multiplication. Then you can follow the multiplication steps above.

So to review $\frac{4}{5} \div \frac{3}{7} = \frac{4}{5} \cdot \frac{7}{3}$

Let's look at an example for dividing rational expressions:

$\frac{x+3}{4} \div \frac{2x+6}{3}$ First, let's change this into a multiplication problem.

$\frac{x+3}{4} \cdot \frac{3}{2x+6}$ Now we can follow the rules for multiplication. Let's factor anything we can in these two rational expressions. The first one cannot be factored at all, but the denominator in the second one can be.

$\frac{x+3}{4} \cdot \frac{3}{2(x+3)}$ Now we can see that there is a common (x+3) that can be cancelled out from both the numerator and denominator.

$\frac{1}{4} \cdot \frac{3}{2} = \frac{3}{8}$ Lastly, we multiply across and simplify, if possible, to get our final answer.

We can also see rational expressions in *rational functions.* A rational function is nothing more than a quotient of two polynomial functions. In a rational function, we can find the zeros or roots of the function by setting the numerator equal to zero and solving. We can also find any vertical asymptotes or holes by setting the denominator equal to zero and solving. Once we find these values, we are able to graph the rational function.

For example, you can graph the rational function $y = \frac{x+2}{x-3}$.

The zeros or roots are found by setting the numerator equal to 0 and solving.

$x + 2 = 0$ So the root of this function will be found at x= -2.

A vertical asymptote would be found by setting the denominator equal to 0 and solving.

$x - 3 = 0$ So at x=3, this function will have an asymptote.

Linear Equations and Inequalities

Linear equations are equations where the exponent of the variable is one. They can have one variable or several variables. One-variable linear equations are solved for the variable given in the equation. These types of equations can require one, two, or multiple steps to solve. Let's look at an example of a multi-step equation with a single variable on both sides.

$3x - 10 + 4x = 5x - 12$	Our goal is to solve for the variable, x. The first step is to look at each side of the equation separately and see if there are any like terms that can be combined. On the right side, there is nothing to combine, but on the left side we can combine the 3x and 4x.
$7x - 10 = 5x - 12$	Next, we want to isolate the variable on one side of the equation. In other words, we want to put all of the terms with the variable on one side and all of the other terms on the other side of the equation.
$7x - 10 = 5x - 12$	To move a term to the other side, take the opposite sign of the term
$-5x \quad -5x$	and add/subtract that from both sides.
$2x - 10 = -12$	If we move the 5x over to the left side of the equation to be with the 7x, we can sum up those terms to get 2x on the left side. On the right side, the two 5x terms cancel, and we are left with -12.
$2x = -2$	Next, we move the -10 to the right side of the equation by adding 10 to both sides to isolate the variable term.

$x = -1$ Finally, we solve for x by dividing both sides of the equation by 2.

Linear equations with one variable typically have one solution as shown in the example above. However, there are instances where there can be no solution or an infinite number of solutions. Let's look at these special cases.

If there is no value for the variable that makes the equation true, then there is no solution. For example:

$$2x + 10 = 2x - 5$$

If we tried to solve this equation for the variable, x, we would find out that the variable actually cancels out, and we would be left with $10 = -5$.

This statement is not true. Therefore, there is no solution to the linear equation.

Conversely, if any value for the variable makes the equation true, then there is an infinite number of solutions. Take a look at the following example:

$$4x - 2 = 2(2x - 1)$$

If we tried to solve this equation for the variable, x, we would see that the variable cancels out, and we would be left with $-2 = -2$.

This statement is true. Therefore, any value that we substitute for x will result in a true statement, and that means that there are an infinite number of solutions for this linear equation. If you try it out, you will see that whatever number you plug in for x on the left side will give you the same thing on the right side. So the solutions are infinite.

Now, let's look at linear equations with two variables:

These types of linear equations describe the relationship between two variables, x and y. A solution to a linear equation with two variables is an ordered pair (x, y) that makes the equation true. If we graph all solutions of a linear equation as points on the coordinate plane, they will form a straight line.

There are certain characteristics of a linear equation that can be determined once we have the solution points and graph:

The steepness of the line, called the *slope,* can be determined given two points on the line. The slope is always constant for any linear equation. It is also referred to as "the change in y over the change in x," or the "rise over run." We use the letter m to denote the slope.

$$m = \frac{\Delta y}{\Delta x} = \frac{rise}{run} = \frac{y_2 - y_1}{x_2 - x_1} \text{ where } (x_1, y_1) \text{ and } (x_2, y_2) \text{ are points on the line.}$$

If the line is a horizontal line, then the slope is 0. If the line is a vertical line, then the slope is undefined. Slopes can be either negative or positive. If the line is going up and to the right, the slope is positive. If the slope is going down and to the right, then the slope is negative.

The *y-intercept* is the point where the line crosses the y-axis, and the *x-intercept* is the point where the line crosses the x-axis.

There are a few different ways to write a linear equation. The different forms are listed below, starting with the most widely used:

- Slope-Intercept Form (use this form when you know the slope and y-intercept):

  $y = mx + b$, where $m = $ slope and $b = y$ intercept

- Standard Form :

  $Ax + By = C$, where A, B, C are constants; slope $= \frac{A}{B}$; y intercept $= \frac{C}{B}$

- Point-Slope Form (use this form when you know a point on the line and the slope):

  $y - y_1 = m(x - x_1)$, where $m = $ slope and (x_1, y_1) is a point on the line

- Two-Point Form (use this form if you know two points on the line):

  $\frac{y - y_1}{x - x_1} = \frac{y_2 - y_1}{x_2 - x_1}$, where (x_1, y_1) and (x_2, y_2) are points on the line

- Intercept Form (use this form if you know the x and y intercepts):

  $\frac{x}{x_1} + \frac{y}{y_1} = 1$, where $x_1 = x$ intercept and $y_1 = y$ intercept

The information that you are given in a problem will help you to decide which is the best form to use.

There are two special cases of linear equations that are important to know:

$y = mx$, where y is directly proportional to x ; $y = \frac{m}{x}$, where y is inversely proportional to x

Now let's look at linear inequalities. Linear inequalities are similar to equations, except that there is an inequality symbol in place of the equal sign, meaning that the two sides are not equal. There are linear inequalities in one and two variables. We are going to focus on inequalities in one variable.

There are four inequality symbols that can be used. They are:

- $>$ *greater than,*
- $<$ *less than,*
- $\geq$ *greater than or equal to,*
- $\leq$ *less than or equal to*

Just like one-variable equations, the solution for the one-variable inequality is the value of the variable that makes the statement true. Typically, there is a range of values that are solutions for an inequality, not just one value like there is for an equation. Let's look at an example:

$4x + 10 \leq 22$ We solve an inequality the same way that we solve an equation. We need to isolate and solve for the variable, x.

$4x \leq 12$ First, we subtract 10 from both sides to isolate x.

$x \leq 3$ Next, we divide both sides of the inequality by 4 to solve for x.

The solution is $x \leq 3$. This means that any number less than or equal to 3 makes the inequality true. So 3, 2, 1, 0, ... and so forth are all solutions, all the way down to $-\infty$.

When we give the solution for an inequality, it is also common to show the solution on a number line. For the example we just did, that would look like this:

We draw a closed circle at 3 because the solution includes 3, and we shade in the number line to the left of 3 all the way to the arrow to show that the solution includes everything less than 3, also.

If the solution is either > or < and doesn't include the equal-to bar, then we would use an open circle instead of a closed circle on the number in the solution.

Let's look at one more example:

$-5x - 7 < 18$ First, add 7 to both sides of the inequality to isolate the variable, x.

$-5x < 25$ Next, divide both sides of the inequality by -5 to solve for x.

                 Please note that there is a special rule when multiplying or dividing by a negative number in an inequality. If you do this, then you must reverse, or "flip," the inequality symbol. So in this example, when we divide by -5 the inequality changes from < to >.

$x > -5$ This is the solution. Now let's graph it.

In this example, since the solution is everything greater than -5 but not including -5, we place an open circle at -5. Then we shade in everything greater than or to the right of -5 all the way to the right arrow.

Just as there were special cases for solutions of linear equations, the same holds true for linear inequalities. There will be no solution if the variable cancels out and the inequality is reduced to something such as $-8 > 3$. This statement cannot be true because -8 is never greater than 3, so that means that this inequality has no solution. Similarly, if the variable cancels out but you are left with an inequality such as $2 < 9$, which is a true statement, the solution is all real numbers.

Systems of Equations

Two or more linear equations consisting of the same variables, such as x and y, form a system of linear equations. We can solve these equations together to find the solution to the system. The solution is a pair of numbers that can be written as an ordered pair. If we graphed both of the linear equations on the same coordinate plane, the point where the two lines intersected would be the solution to the system of equations.

There are two methods that are commonly used to solve a system of equations. These methods are known as *substitution* and *elimination* (also known as *combination*).

Let's look at each method, starting with substitution:

When using substitution to solve a system of equations, there are a number of steps:

Step 1: Solve one of the equations for one of its variables. Try to choose the variable that is easiest to solve for.

Step 2: Substitute the answer from Step 1 into the other equation and solve for the other variable.

Step 3: Substitute the value for the variable obtained in Step 2 into the answer from Step 1 to solve for the remaining variable.

Step 4: Check the solution by plugging the values for the variables into each of the original equations and solving.

Step 5: Write the solution as an ordered pair (x, y).

Let's look at an example:

Eq. 1: $3x + 5y = 25$

Eq. 2: $x - 2y = -10$

Step 1: The easiest variable to solve for here is x because it is already by itself. So let's solve Eq. 2 for x.

$$x = -10 + 2y$$

Step 2: Substitute $x = -10 + 2y$ into Eq. 1 for x and solve for y.

$$3(-10 + 2y) + 5y = 25$$
$$-30 + 6y + 5y = 25$$
$$-30 + 11y = 25$$
$$11y = 55$$
$$y = 5$$

Step 3: Substitute $y = 5$ into the revised equation from Step 1 and solve for x.

$$x = -10 + 2(5) = -10 + 10 = 0$$

Step 4: Check the solution for x and y by plugging both values into the original equations. By plugging x=0 and y=5 into equation 1, we get 25=25, which is true. Equation 2 gives us -10 = -10, which is also true. Now we know that our answer is correct.

Step 5: The solution is (0, 5).

Let's look at the second method, elimination, or combination:

Similar to substitution, elimination (or combination) requires several steps to solve a system of equations.

Step 1: Arrange the equations with like terms in columns.

Step 2: If necessary, multiply one or both of the equations by numbers that will make the coefficients of one of the variables opposites, so that they can cancel out.

Step 3: Add the revised equations from Step 2 to solve for the remaining variable that isn't canceled out.

Step 4: Plug the answer from Step 3 into either of the original equations to solve for the other variable.

Step 5: Check the solution for both variables by plugging them into each of the original equations.

Step 6: Write the solution as an ordered pair (x, y).

Let's look at an example:

Eq. 1: $2x - 3y = 4$

Eq. 2: $-4x + 5y = -8$

Step 1: The equations are already arranged in the correct order, with the x terms, y terms, and coefficients in their own columns.

Step 2: Multiply Eq. 1 by 2 in order to change the coefficient of x to 4 so that x can be cancelled out:

$$2(2x - 3y = 4)$$

$$4x - 6y = 8$$

Step 3: Add the revised equation from Step 2 with Eq. 2 to cancel out the x and solve for y:

$$4x - 6y = 8$$
$$+ \quad -4x + 5y = -8$$

$-1y = 0$, so $y = 0$

Step 4: Plug y=0 into either of the original equations and solve for x:

$$2x - 3(0) = 4$$
$$2x = 4$$
$$x = 2$$

Step 5: Check the solutions for x and y by plugging them into each of the original equations.

Step 6: The solution is (2, 0).

Proportions

Proportions are equations that state that two ratios are equal. Remember, a ratio is the comparison of one quantity to another. Ratios are expressed as fractions. Proportions can be solved for the unknown quantity or variable in the equation.

An example of a proportion is $\frac{5}{6} = \frac{10}{4x}$

We use something called the *cross product property* to solve proportions.

This property states: *If* $\frac{a}{b} = \frac{c}{d}$ *, then* $a \cdot d = b \cdot c$.

We cross multiply to eliminate the fractions and help simplify the equation so we can solve for the variable.

Let's look at an example:

$\frac{2}{x} = \frac{16}{40}$ Use the cross product property to cross multiply. This becomes:

$2 \cdot 40 = x \cdot 16$ Simplify both sides.

$80 = 16x$ Divide both sides of the equation by 16 to get the solution:

$$5 = x \text{ or } x = 5$$

Let's look at one more example:

$\frac{r+4}{3} = \frac{r}{5}$ Cross multiply.

$(r+4)5 = 3r$ Simplify both sides.

$5r + 20 = 3r$ Combine like terms and isolate r on the left side of the equation.

$2r = -20$ Divide both sides of the equation by 2 to solve for r:

$$r = -10$$

There is another way to solve some of the simpler proportions, such as our first example. Let's look at that again: $\frac{2}{x} = \frac{16}{40}$

Since we know that the two ratios or fractions are equal, we can compare the two numerators and the two denominators to see if we can identify a pattern for one of them. In this proportion, let's compare the numerators since the denominators include the variable. If we think about 16, what do we need to do to 16 to change it to 2 (think multiplication or division)? We would need to divide 16 by 8 to make it 2.

Therefore, we can do the same thing to the denominator. If we divide 40 by 8, what do we get? 5.

And that was the same answer we got by solving it using the cross product property above. So in certain cases there will be more than one method for solving a proportion. Choose the method that is easiest for you.

Word Problems and Applications

Word problems are math problems that are written out in words and describe a realistic scenario or event that needs to be solved. When working through word problems, there are a series of steps that should be used to help solve them. They are as follows:

Step 1: Read the problem carefully.

Step 2: Reread the problem and underline any important numbers or words that you will need to solve the problem.

Step 3: Determine what the problem is asking for and circle it.

Step 4: Draw a picture or diagram to help you better understand the problem and what it's asking for.

Step 5: Determine the best equation to solve the problem and write it out.

Step 6: Solve the equation.

Step 7: Check your answer. Do you have the appropriate units with your answer? Does your answer make sense?

There are many types of word problems relating to the many different topics in math. We will look at three of the most common types.

1. **Distance Word Problems:**

Distance word problems are seen all the time in everyday life. Let's look at an example:

John and Andrea are driving from their home in North Carolina to their friend Mark's house in Georgia. If the trip is 480 miles, how long will it take them if they drive at an average speed of 60 miles per hour?

With any distance problem, you should use the distance formula, which is a linear equation:

$d = rt$, where $d = distance$, $r = rate\ or\ speed$, and $t = time$

Using the problem-solving steps, we would want to underline the following important information: 480 miles and 60 mph. We would also want to circle "how long will it take" because that is what the problem is asking for. We already said that we will use the distance formula, so let's set it up.

In this problem, $d = 480$ and $r = 60$ and t is what we are looking for.

$480 = 60t$ — Divide both sides of the equation by 60 to solve for t.

$t = 8\ hours$ — It will take 8 hours for John and Andrea to get to their friend Mark's house.

2. Percent Word Problems

Alan took a math test and answered 78% of the problems correctly. If there were 45 problems on the test, how many problems did he get correct?

We can use a proportion to solve this problem.

When reading through the problem, we should underline "answered 78% of the problems correctly" and "45 problems on the test." We should also circle "how many problems did he get correct," as this is what is being asked.

Now let's set up our proportion. Hint: Sometimes it helps to use words first before inserting the numbers.

$$Percent\ Correct = \frac{number\ of\ problems\ correct}{total\ number\ of\ problems} \rightarrow \frac{78}{100} = \frac{x}{45}$$

$78 \cdot 45 = 100 \cdot x$ Cross multiply to solve for x.

$3510 = 100x$ Divide both sides of the equation by 100 to solve for x.

$x = 35.1\ problems$ Alan got 35 problems correct on his math test.

3. Systems of Equations Word Problems

Sam bought 4 shirts and 3 pairs of pants for $181. Jenna bought 1 shirt and 2 pairs of pants for $94. If all of the shirts and pants are the same price, what is the price for each shirt and each pair of pants?

With this type of problem, we are going to set up two equations and solve them together as a system of equations. When reading through the problem, 4 shirts, 3 pairs of pants, $181, 1 shirt, 2 pairs of pants, and $94 should be underlined. And "what is the price for each shirt and each pair of pants" should be circled, as this is what the problem is asking for.

Let's set up the equations. There will be one equation for Sam and one equation for Jenna. The unknowns we are looking for are shirts and pants, so let's assign variables to these: x=shirts and y=pants.

Eq. 1 for Sam: $4x + 3y = 181$

Eq. 2 for Jenna: $1x + 2y = 94$

We can use either substitution or elimination to solve this system. Let's use substitution.

$1x + 2y = 94$ Solve Eq. 2 for x.

$x = 94 - 2y$ Substitute this new revised equation into Eq. 1.

$4(94 - 2y) + 3y = 181$ Solve for y:

$$376 - 8y + 3y = 181$$

$$376 - 5y = 181$$

$$-5y = -195$$

$y = \$39$ Substitute the value for y into the revised Eq.2 and solve for x:

$$x = 94 - 2y$$

$$x = 94 - 2(39)$$

$x = 94 - 78 = \$16$ Each shirt cost $16, and each pair of pants cost $39.

Before You Start the Practice Questions

On the next page, you will begin a Math Knowledge practice test. Set a timer for 22 minutes before you start this practice test. Giving yourself 22 minutes will give you an authentic feel for how long you have to finish this portion of the AFOQT. Like the official test, this practice test has exactly 25 items. Set your timer, turn the page, and begin.

Math Knowledge Practice Test Questions

(22 minutes)

1. $\frac{2}{7} + \frac{1}{42} - \frac{3}{14} = ?$

    a. 0

    b. $\frac{5}{7}$

    c. $\frac{2}{21}$

    d. $\frac{3}{35}$

    e. $\frac{5}{35}$

2. Solve for x and y, given: $3x + y = 3$ and $7x + 2y = 1$

    a. $x = -5, y = 18$

    b. $x = 3, y = 1$

    c. $x = 10, y = 3$

    d. $x = 7, y = 2$

    e. $x = 11, y = 3$

3. $\sqrt[3]{64} - 2 = ?$

    a. 20

    b. 190

    c. 14

    d. 2

    e. 7

4. The factors of $x^2 - 3x + 2$ are:

    a. $(x-3)(x+2)$

    b. $(x-2)(x-1)$

    c. $(x+2)(x+1)$

    d. $(x+2)(x-1)$

    e. $(x+3)(x-1)$

5. The right triangle ABC has A = 36°, what are the values of B and C?

    a. 90 and 234

    b. 45 and 90

    c. 36 and 90

    d. 90 and 54

    e. 90 and 290

6. If you have a square vegetable garden that is 225 ft² in area, how much fence would you need to put around the garden?

    a. 225 feet

    b. 30 feet

    c. 56.25 feet

    d. 60 feet

    e. 85 feet

7. $\frac{3}{5} \div \frac{12}{25} = ?$

    a. $\frac{5}{4}$

    b. $\frac{4}{5}$

    c. $\frac{36}{125}$

    d. $\frac{3}{25}$

    e. $\frac{7}{5}$

8. Sam and his church group are going to an amusement park. Adult tickets cost $25 each, children tickets cost $15 each, and the total amount spent for all tickets is $370. If there are 20 total people in the group, how many are children?

    a. 15

    b. 13

    c. 10

    d. 8

    e. 9

9. Given the polynomial equation $b^2 + 4 = 20$, what is one possible solution for b?

    a. $\sqrt{24}$

    b. -2

    c. 16

    d. -4

    e. -5

10. Given a rectangular box that is 3 ft long by 1 ft wide by 1 ft high, what is the volume of the box?

    a. 3 ft²

    b. 5 ft³

    c. 3 ft³

    d. 4 ft³

    e. 5 ft²

11. Given a triangle with an area of 36 cm² and a base equal to 6 cm, what is its height?

    a. 6 cm

    b. 12 cm

    c. 30 cm

    d. 18 cm

    e. 13 cm

12. $\frac{\sqrt[3]{64}}{-2} = ?$

    a. -2

    b. -4

    c. 32

    d. -96

    e. -16

13. Simplify the following polynomial expression: $(x^2 - 4)(x^2 + 4)$

    a. $x^2 - 16$

    b. $x^2 - 4x + 4$

    c. $x^4 - 4$

    d. $x^4 - 16$

    e. $x^4 - 8$

14. Simplify the following expression: $\frac{7}{8} + \frac{3}{10}$

   a. $\frac{47}{40}$
   b. $\frac{10}{18}$
   c. $\frac{10}{80}$
   d. $\frac{21}{80}$
   e. $\frac{11}{80}$

15. If a rectangle has a length that is equal to $\frac{5}{2}$ of its width, and the area is 250 in², what is its width?

   a. 625 in
   b. 25 in
   c. 100 in
   d. 10 in
   e. 125 in

16. What two whole numbers does $\sqrt{40}$ fall in between?

   a. 3 and 4
   b. 7 and 8
   c. 5 and 6
   d. 6 and 7
   e. 4 and 5

17. Solve the following percent problem: 45% of what distance is 135 miles?

   a. 300 miles
   b. 45 miles
   c. 60.75 miles
   d. 3 miles
   e. 50 miles

18. Simplify the following expression: $2\frac{1}{3} - \frac{2}{9}$

   a. $\frac{5}{6}$
   b. $1\frac{1}{9}$
   c. $2\frac{1}{9}$
   d. $\frac{5}{9}$
   e. 2

19. What is $\frac{630}{72}$ rounded to the nearest tenths?

   a. 8.7
   b. 8
   c. 9
   d. 8.8
   e. 8.5

20. Which of the following numbers is largest?

    a. $\frac{2}{3}$

    b. $\sqrt{2}$

    c. 0.34

    d. $\frac{1}{8}$

    e. $\frac{3}{8}$

21. Solve the following equation for x: $3(x+6) = 5x - 20$

    a. 19

    b. 13

    c. 7

    d. -1

    e. 2

22. Choose the equation you would use to find 25% of 120.

    a. $0.25x = 20$

    b. $x = \frac{120}{0.25}$

    c. $x = \frac{0.25}{120}$

    d. $x = 0.25 \cdot 120$

    e. $x = 0.25 \cdot 12$

23. Given $\frac{5a^2b-15a+10b^2}{30ab}$, which of the following is an equivalent expression?

   a. $\frac{a}{6} - \frac{1}{2b} + \frac{b}{3a}$

   b. $\frac{6}{a} - 2b + \frac{3a}{b}$

   c. $\frac{a-1+b}{6-2b+3a}$

   d. $\frac{15a(a+b)}{30ab}$

   e. $a + 1 = 2b - 2$

24. Place the following numbers in decreasing order: $\frac{56}{100}, \sqrt{3}, 0.74, \frac{7}{3}$

   a. $0.74, \frac{7}{3}, \frac{56}{100}, \sqrt{3}$

   b. $\sqrt{3}, 0.74, \frac{56}{100}, \frac{7}{3}$

   c. $\frac{7}{3}, \sqrt{3}, 0.74, \frac{56}{100}$

   d. $\frac{56}{100}, 0.74, \sqrt{3}, \frac{7}{3}$

   e. $\frac{56}{100}, \sqrt{3}, 0.74, \frac{7}{3}$

25. Simplify the following expression: $(4x + 1)(6x - 7)$

   a. $10x^2 - 17x - 8$

   b. $24x^2 - 13x - 7$

   c. $10x^2 - 5x + 7$

   d. $24x^2 - 22x - 7$

   e. $24x^2 - 12x - 7$

Answer Guide to Math Knowledge Practice Test

1. **C.** $\frac{2}{7}+\frac{1}{42}-\frac{3}{14}=?$

In order to solve this, we must find a common denominator so that we can combine the fractions.

If we write out the multiples of all of the denominators:
7: 7,14,21,28,35,42,49,...
14: 14,28,42,56,...
42: 42,84,126,...

We can see that 42 is the least common multiple for all three, and therefore will be our common denominator.

Next, we need to convert all three fractions into equivalent fractions with the denominator 42.

$\frac{2}{7}=\frac{12}{42}$ For the first fraction, we must multiply both the numerator and denominator by 6 to get the equivalent fraction.

$\frac{3}{14}=\frac{9}{42}$ For the third fraction, we must multiply both the numerator and denominator by 3 to get the equivalent fraction.

The second fraction, $\frac{1}{42}$, does not need to be changed.

Now we can combine them by adding the numerators and keeping the denominator of 42:

$\frac{12}{42}+\frac{1}{42}-\frac{9}{42}=\frac{4}{42}=\frac{2}{21}$. Then the final step is to reduce the fraction to get the answer of $\frac{2}{21}$.

2. **A.** Solve for x and y, given: $3x+y=3$ and $7x+2y=1$

This is a system of equations and can be solved using either substitution or elimination (combination).
I am going to solve it by substitution:

First, rearrange the first equation to solve for y: $y=3-3x$.

Next, substitute the equation you just solved for y into the second original equation,
$$7x+2y=1$$
$$7x+2(3-3x)=1$$

Then simplify the left side of the equation: $7x+6-6x=1$, which simplifies to $x+6=1$.

Now solve for x: $x = -5$.

This is only half of the solution. We still need to solve for y.

Take the value you just got for x and plug it into the equation that you got in the first step of the solution: $y = 3 - 3x$ → $y = 3 - 3(-5) = 3 + 15 = 18$.

So the final solution is $x = -5$, $y = 18$.
You can double check this answer by plugging these values into each of the original equations.

3. **D.** $\sqrt[3]{64} - 2 = ?$

The first term is the cube root of 64, which is 4, because $4 \cdot 4 \cdot 4 = 4^3 = 64$.

Therefore, $4 - 2 = 2$.

The final answer is 2.

4. **B.**   The factors of $x^2 - 3x + 2$ are:

To solve this, we have to factor this polynomial. Since it is a second degree polynomial, we can factor it into two factors: $(x \pm ?)(x \pm ?)$.

To find the unknown values or ? in each factor, we need to remember that we are looking for two numbers that multiply together to get 2 and also add up to -3.

Since 1 and 2 are the only factors of two, we know it has to be some combination of those numbers. If we multiply -2 and -1, we get 2, and they also add up to -3.

Therefore, the factors are written as $(x - 2)(x - 1)$.

5. **D.**   The right triangle ABC has $A = 36°$, what are the values of B and C?

There are two important things that you must remember in order to solve this problem: every triangle has angles that add up to 180°, and a right triangle always has a right angle, which is 90°.

So B is 90° and that means that $C = 180 - 90 - 36 = 54°$.

6. **D.** If you have a square vegetable garden that is 225 ft² in area, how much fence would you need to put around the garden?

So this is a word problem. The first thing I would do is underline any important numbers or words that we will need to solve this. I would underline "square" and "225 ft² in area."

Next, I would circle "how much fence would you need to put around the garden" because that is what the problem is asking for.

Now we need to set up an equation to solve it. Actually, two equations. First, we need to use the information that it is a square garden and it is 225 ft² in area. Since it is square, we know that all of the sides have to be the same length.

Remember the formula for the area of a a square is A=(one side)².

So we know that $A = 225 = one\ side^2 \rightarrow one\ side = \sqrt{225} = 15$ ft.

Now that we know the length of each side of the square garden, we just need to find out how much fence we need to go around the garden. The key word here is "around." When you see this word, you should automatically think "around = perimeter."

So we need the fence to go around all four sides, which are each 15 ft long. The perimeter of a square is $P = side + side + side + side = 15 + 15 + 15 + 15 = 4(15) = 60$ ft.

Therefore, we need 60 ft of fence to go around our garden.

7. **A.** $\frac{3}{5} \div \frac{12}{25} = ?$

In order to solve this, we have to change this into a multiplication problem by changing the division to multiplication and flipping the second fraction.

So the equation becomes $\frac{3}{5} \cdot \frac{25}{12} = ?$

We can either simplify first and then multiply, or multiply first and then simplify. Usually it is easiest to try to simplify first and then multiply.

So if we do this, we can reduce the 3 and 12 to 1 and 4, and we can reduce the 5 and 25 to 1 and 5.

Then we have $\frac{1}{1} \cdot \frac{5}{4} = \frac{5}{4}$. The final answer is $\frac{5}{4}$.

8. **B.** Sam and his church group are going to an amusement park. Adult tickets cost $25 each, children tickets cost $15 each, and the total amount spent for all tickets is $370. If there are 20 total people in the group, how many are children?

This is another word problem, so we need to underline the important information from the problem that we will need to solve it. We should underline adult tickets cost $25, children tickets cost $15, total amount spent for all tickets is $370, and 20 total people in the group.

Next, circle "how many are children" because this is what is being asked in the problem.

This will require two equations and, therefore, will be a system of equations to solve.

The first equation will use the information that the total number of people is 20. We will assign variables to the adults and the children: $adults = x$ and $children = y$.

So for the first equation, we can write $x + y = 20$, which says that the adults plus children equals 20 people.

For the second equation we are going to utilize how much each ticket costs and the total amount spent.

$25x + 15y = 370$, which says that for x adults, we need to multiply that by the cost of 25 for each adult ticket, and similarly for y children we need to multiply that by the cost of 15 for each child ticket. Then if we add those together, we should get the total amount of money spent, which is 370.

Now we have our system of equations: $x + y = 20$,

$$25x + 15y = 370$$

Let's use substitution to solve this:

First, we need to rearrange Eq.1 to solve for either x or y. Let's solve for x: $x = 20 - y$

Next, substitute that new equation into Eq.2 above: $25(20 - y) + 15y = 370$
Then solve for y:

$$500 - 25y + 15y = 370$$
$$500 - 10y = 370$$
$$-10y = -130$$

$y = 13 \ children$

We could stop here, because that's all the problem asked for, but let's finish solving the system and find out how many adults are also in the group.

Take the value for y and plug it into the rearranged equation for x in Step 1:

$$x = 20 - y = 20 - 13 = 7 \ adults$$

There are 13 children and 7 adults in the group.

9. **D.** Given the polynomial equation $b^2 + 4 = 20$, what is one possible solution for b?

In this problem we need to solve for b. So we need to isolate b on the left side of the equation:

$$b^2 + 4 = 20$$
$$b^2 = 16$$
$$b = \pm 4$$

10. **C.** Given a rectangular box that is 3 ft long by 1 ft wide by 1 ft high, what is the volume of the box?

The formula for the volume is $V = Length \cdot Width \cdot Height$.

Therefore, in this problem $V = 3 \cdot 1 \cdot 1 = 3$. What units should we use?

We are multiplying $t \cdot feet \cdot feet$, so we should have ft³, or cubic feet.

11. **B.** Given a triangle with an area of 36 cm² and a base equal to 6 cm, what is its height?

For this problem, you need to remember the formula for the area of a triangle:

$A = \frac{1}{2}bh$, where $b = base$ and $h = height$

In this problem we are given A and b, and we need to solve for h. Let's plug in the values we know and solve:

$$36 = \frac{1}{2}(6)h \;\;\rightarrow\;\; 36 = 3h \;\;\rightarrow\;\; h = 12\; cm$$

12. **A.** $\frac{\sqrt[3]{64}}{-2} = ?$

$\sqrt[3]{64} = 4$ because $4 \cdot 4 \cdot 4 = 4^3 = 64$.

Therefore, $\frac{\sqrt[3]{64}}{-2} = \frac{4}{-2} = -2$.

13. D.   Simplify the following polynomial expression: $(x^2 - 4)(x^2 + 4)$

There are two ways to work through this problem. The first way is to use FOIL to simplify, and the second way is to recognize that the expression given is actually a difference of squares and simplify according to that.

Let's solve by FOIL first:

$$(x^2 - 4)(x^2 + 4) = x^2 \cdot x^2 + 4 \cdot x^2 - 4 \cdot x^2 - 4 \cdot 4 = x^4 + 4x^2 - 4x^2 - 16 = x^4 - 16$$

If we recognize from the start that $(x^2 - 4)(x^2 + 4)$ is the factored form of a difference of squares, we can easily solve this. Remember that a difference of squares is:

$$a^2 - b^2 = (a - b)(a + b).$$

Therefore, for this expression, $a = x^2$ and $b = 4$, so our answer would be:

$$(x^2)^2 - 4^2 = x^4 - 16.$$

14. A.   Simplify the following expression: $\frac{7}{8} + \frac{3}{10}$

For this problem we need to find a common denominator for the fractions to be able to add them.

Write out the multiples of both 8 and 10:

8: 8, 16, 24, 32, 40, 48, 56, ...

10: 10, 20, 30, 40, 50, ...

We see that the least common multiple between them is 40, so that is our LCD.

$\frac{7}{8} = \frac{35}{40}$ and $\frac{3}{10} = \frac{12}{40}$

Now we can add the equivalent fractions with the common denominators:

$$\frac{35}{40} + \frac{12}{40} = \frac{47}{40}$$

15. D. If a rectangle has a length that is equal to $\frac{5}{2}$ of its width, and the area is 250 in², what is its width?

Remember that the area of a rectangle is $A = L \cdot W$, where $L = length$ and $W = width$.

We know the area and we know that the length is equal to $\frac{5}{2}$ of its width.

So we can actually write that as $L = \frac{5}{2}W$.

Then we can plug in what we know to the area equation: $250 = \frac{5}{2}W \cdot W = \frac{5}{2}W^2$.

If we then multiply both sides of the equation by $\frac{2}{5}$ (the reciprocal of $\frac{5}{2}$) in order to cancel the fraction out, we end up with: $\frac{500}{5} = W^2 \rightarrow 100 = W^2 \rightarrow W = \sqrt{100} \rightarrow W = 10\ in.$

16. D. What two whole numbers does $\sqrt{40}$ fall in between?

If we look for two perfect squares, one on either side of 40, we can figure this out.

The two perfect squares are 36 and 49: $36 = 6^2\ and\ 49 = 7^2$.

Therefore, $\sqrt{36} = 6\ and\ \sqrt{49} = 7$, which means that $\sqrt{40}$ has to be in between 6 and 7.

17. A. Solve the following percent problem: 45% of what distance is 135 miles?

We can use a proportion to solve this problem.

$45\% = \frac{45}{100}$, so we write the proportion like this: $\frac{45}{100} = \frac{135}{x}$.

Solve the proportion by cross-multiplying: $100 \cdot 135 = 45 \cdot x \rightarrow 13500 = 45x$.

If we divide by 45 on both sides of the equation to solve for x, we get $x = 300\ miles$.

18. C. Simplify the following expression: $2\frac{1}{3} - \frac{2}{9}$

In order to solve this one, we need to do two things: Convert the first mixed number into an improper fraction and then find a common denominator between them:

$2\frac{1}{3} = \frac{7}{3}$

If you don't remember how to do this, you multiply the whole number part of the mixed number by the denominator and then add that to the numerator to get the new numerator for the improper fraction. So $2 \cdot 3 + 1 = 7$. We keep the denominator in the mixed number for the denominator of the improper fraction, making the improper fraction equal to $\frac{7}{3}$.

Now we have $\frac{7}{3} - \frac{2}{9}$.

It should be easy to see that the common denominator between these two fractions is 9. Therefore, we only have to change the first fraction: $\frac{7}{3} = \frac{21}{9}$

Now, let's solve: $\frac{21}{9} - \frac{2}{9} = \frac{19}{9} = 2\frac{1}{9}$ (If you don't remember how to convert from an improper fraction back to a mixed number, divide the 9 into the 19 and you get 2 with a remainder of 1. So the 2 becomes the whole number part of the mixed number, and the 1 becomes the numerator of the fraction part).

19. **D.** What is $\frac{630}{72}$ rounded to the nearest tenths?

If we divide 630 by 72, we get an answer of 8.75.

The problem asks us to round to the nearest tenths. To round, we look at the digit to the right of the tenths position and see if it is < 5 or ≥ 5. In this case, the tenths position is 7, so we look at the digit to the right of the 7, which is 5. Since it is ≥ 5, we round our tenths digit up to 8.

Our final answer is 8.8.

20. **B.** Which of the following numbers is largest?

The numbers we are given to choose from are. $\frac{2}{3}, \sqrt{2}, 0.34, \frac{1}{8}$.

To figure this out, we need to convert all of these numbers into the same form to be able to compare them. We have fractions, roots, and decimals. If we convert them all to decimals, we will be able to easily see which is the largest:

$$\frac{2}{3} = 0.67, \sqrt{2} = 1.41, 0.34 \text{ stays the same}, \frac{1}{8} = 0.125$$

By comparing these decimals now, I can see that 1.41 is the largest, which is $\sqrt{2}$.

21. **A.** Solve the following equation for x: $3(x + 6) = 5x - 20$

In this problem, we must simplify each side of the equation first, then isolate the variable to solve:

$$3(x + 6) = 5x - 20 \quad \rightarrow \quad 3x + 18 = 5x - 20$$

We need to isolate x, so let's move the -20 over to the left side of the equation to be with the 18 by adding 20 to both sides, and then move the 3x to the other side to be with the 5x by subtracting 3x from both sides. This becomes $38 = 2x$.

Divide both sides of the equation by 2 to get the answer x=19.

22. **D.** Choose the equation you would use to find 25% of 120.

For this percent problem, you need to set up a proportion. So let's do that first:

Remember that $25\% = \frac{25}{100} = 0.25$, so $0.25 = \frac{x}{120}$.

Now, to isolate and solve for x, we need to multiply both sides of the equation by 120. This gives us: $0.25 \cdot 120 = x$.

23. **A.** Given $\frac{5a^2b - 15a + 10b^2}{30ab}$, which of the following is an equivalent expression?

Whenever we have a fraction with multiple terms in the numerator and one term in the denominator, we can split this fraction into separate fractions by placing each term in the numerator over the denominator and keeping the existing operations between them.

So for this expression, we have three terms in the numerator. We can separate this out into three fractions separated by their respective operations.

$\frac{5a^2b}{30ab} - \frac{15a}{30ab} + \frac{10b^2}{30ab}$ Now we just have to simplify each of the terms by reducing them:

For the first term, $\frac{5a^2b}{30ab} = \frac{a}{6}$.

For the second term, $\frac{15a}{30ab} = \frac{1}{2b}$.

For the third term, $\frac{10b^2}{30ab} = \frac{b}{3a}$.

Putting all three terms back together with the correct operations makes the expression $\frac{a}{6} - \frac{1}{2b} + \frac{b}{3a}$.

24. C. Place the following numbers in decreasing order: $\frac{56}{100}, \sqrt{3}, 0.74, \frac{7}{3}$.

For this problem, in order to be able to compare them, we need to convert them all into the same form. We have fractions, decimals, and roots. Let's convert all of them into decimals:

$$\frac{56}{100} = 0.56, \sqrt{3} = 1.73, 0.74 \text{ stays the same}, \frac{7}{3} = 2.33$$

Putting them into decreasing order means ordering them from the largest number down to the smallest number.

Doing this, we get 2.33 , 1.73 , 0.74 , 0.56.

If we then put the numbers back into their original forms, they become: $\frac{7}{3}, \sqrt{3}, 0.74, \frac{56}{100}$.

25. D. Simplify the following expression: $(4x + 1)(6x - 7)$

Use FOIL to simplify this expression: $4x \cdot 6x - 7 \cdot 4x + 1 \cdot 6x - 1 \cdot 7$.

This becomes: $24x^2 - 28x + 6x - 7$.

Finally, combine any like terms. We can combine the -28x and the 6x to get -22x.

The final answer is $24x^2 - 22x - 7$.

READING COMPREHENSION

This portion of the AFOQT is much like other standardized tests. You will be given a passage to read and asked questions based on the text. Each answer will be selected from a list of five possible options. Not only will you need to understand what you are reading, but time will be of the essence.

This portion will assess your ability to process and understand large amounts of written material. These passages will be on a reading level common in scientific and academic publications. Most magazines, website pages, and articles are written at a level much lower than what you will encounter in this section.

Although some of the writing may be technical, no prior knowledge of any of the material will be expected. You won't need any previous specialized knowledge to answer the questions; you will only need the given text.

Among other skills, your ability to locate details, identify the main idea, and draw conclusions in a timely manner will be tested. Before we jump into the practice test, let's explore some different aspects of reading comprehension.

Big Picture and Main Points

When reading a passage of text, it is important to catch the big picture of the passage. "What is the gist of the text?" is a good question to ask yourself. The passage can be broken into two parts: topic(s) and main idea(s).

The topic is the general subject of the passage. Think of a book. A book typically has an overall theme. Maybe the book is about biology or business. The book is made of chapters that go into more detail, yet the chapters still operate within the big picture or topic of the book. On a much smaller scale, paragraphs and passages have topics as well. A paragraph will have a topic (or topics). The topic is the overall subject, and everything within the passage contributes to the topic.

If the topic is the house of the paragraph, the main ideas are like rooms in that house. The main ideas exists inside the framework of the topic(s). The main ideas contribute to the bigger picture (topic).

The main ideas consist of details. Think of the topics as the "what" and the main ideas as the "why" or "how." If someone were to ask you about a particular book you were reading, you may respond with, "It's a book about test taking (topic). It describes how diet and exercise boost test scores (main ideas)."

Since time is limited, the art of skimming is a very important skill to have in this portion of the test. Oftentimes, simply skimming the passage can be an efficient and effective way to determine the topic(s) and main idea(s). Later we will discuss topic sentences and how they can be used to quickly identify the topic.

To have a more practical understanding of how this works during a reading comprehension test, let's dissect a paragraph and identify the topics and main ideas. In the passage below, words or sentences relating to the topic are in **bold** and words or sentences relating to the main ideas are underlined.

> **Professor Langley called his machine the *Aerodrome*, and by October, 1903, the plane was ready for its test flight, with Manley to guide it**. The *Aerodrome* was to be launched from a catapulting platform built on the roof of a houseboat. The houseboat was anchored on the Potomac River near Washington. As it left the platform the machine crashed into the river, and the trial was a dismal failure. The newspapers and the public ridiculed Langley, but he and Manley, who was unhurt in the crash, repaired the machine for another trial. This test took place on December 8, 1903, and again the *Aerodrome* crashed into the river. Manley once more escaped injury, but Langley and the government were abused by the public for wasting money. Langley was out of money himself, and the government could not furnish funds for further trials, so the experiments were ended. The professor, discouraged and brokenhearted, gave up.

The topic of the paragraph is the test flight of the Aerodrome. A few main ideas exist in the passage above. First, the test flights were a failure. Second, failures led to public ridicule. Third, the failures and public ridicule led to giving up. Identifying the topic(s) and main ideas gives you a better understanding of the overall message of the passage. Grasping the big picture brings clarity to the text. If you are asked a question about the paragraph, you will have a general understanding of what the paragraph means and have a better chance of accurately answering the questions.

Small Picture and Details

Supporting details give more information to expound upon the topics and main ideas. If the topic is the house and the main ideas are the rooms, the supporting details are the wood and nails. The supporting details give specifics. Let's take a look at another passage about aviation. Many of the supporting details are in **bold**.

> Out in **Dayton, Ohio**, there were two **small** brothers, who dreamed of flying like birds through the air. Their dreams were heightened by a small toy given to them by their father, **the pastor of a local church**. This toy was to lead to an idea which had a profound effect on the world. You would probably call it a flying propeller. **It consisted of a wooden propeller which**

slipped over a notched stick. By placing a finger against the propeller and rapidly pushing it up the notched stick, the propeller was made to whirl up off the end of the stick and fly into the air. The brothers, young as they were, never quite forgot this little toy as they continued to dream of flying like birds through the air.

The topic of this passage is "Two brothers were inspired to fly." A couple main ideas are in this passage. First, two brothers had their dream of flying heightened by a toy from their father. Second, the toy gave them ideas about how to create a propeller. The big picture about the two dreamy boys getting a toy from their father is supported by several details. The passage notes they were in Dayton, Ohio. The passage mentions that they are small and that their father was the pastor of the local church. The text gives clear details about how the toy was made and how it was used. Without the supporting details, the reader could still get the general idea of the passage, but the supporting details give some specifics to the main ideas.

During the test, you will be asked questions relating to supporting details. Even if you grasp the overall meaning of the passage, you may have to skim the text for details to answer certain questions. If the question is, "What material was the toy made from?" the answer would be "wood." The material of the toy does not change the meaning of the passage, but it is a detail that is included in the text. Some questions will focus on the big picture ideas, and some questions will focus on the small picture details.

The First and Last Sentence

Many passages contain what is known as a *topic sentence.* Often you can spot the topic of the paragraph simply by looking at the first sentence (although occasionally, the topic sentence is found later in a paragraph). Sometimes this is not the case, so do not expect every paragraph to have a topic sentence. This is just another tool to help you skim paragraphs for relevant data to correctly answer test questions.

The last sentence of a paragraph is often a summary sentence. It is a sentence that wraps up the paragraph. Keep in mind that the last sentence is not necessarily a summary sentence. Among other possibilities, the last sentence is often a transition sentence to the next paragraph. This is merely a technique to keep in your back pocket to quickly spot topics and conclusions of the given text.

Let's take another look at the paragraph in the previous section and analyze the first and last sentences. The first and last sentences are in **bold**.

Out in Dayton, Ohio, there were two small brothers, who dreamed of flying like birds through the air. *Their dreams were heightened by a small toy given to them by their father*, the pastor of a local church. This toy was to lead to an idea which had a profound effect on the world. You would probably call it a flying propeller. It consisted of a wooden propeller which slipped over a notched stick. By placing a finger against the propeller and rapidly pushing it up the notched stick, the propeller was made to whirl up off the end of the stick and fly into the air. **The brothers, young as they were, never quite forgot this little toy as they continued to dream of flying like birds through the air.**

The first sentence in **bold** sets up the paragraph to be about two brothers who dreamed of flying. The last sentence in **bold** concludes that a toy would somehow contribute to this dream. By merely looking at the first and last sentences, the reader gets a fairly good idea of what the paragraph is about. If something is unclear, the reader simply has to look at the first half of the second sentence to get a very accurate understanding of the passage. The second sentence (in italics) clarifies any confusion about what's going on. The rest of the paragraph is full of supporting details about how this toy inspired them, but the reader can extract quite a bit of information by reading the first and last sentences and skimming the rest.

Realistically, not all text you will encounter will have a neatly structured order of sentences. It would be much easier if every paragraph you encountered followed the same pattern, but that won't happen. Many times, it will be hard to skim the text for topic or summary sentences, because they are not contained in the text. In times where no topic or summary sentence exists, the reader has to dive deeper into the paragraph and read further for a clearer understanding.

Inferences

Making an inference involves using reason and evidence to draw a conclusion. Inferences are based on implied information that is not stated outright in the text. Skimming the text is usually an incomplete approach to making an inference. Typically, inferences are only accurately formed by looking at the details. Readers can easily draw incorrect conclusions by relying too much on their own imaginations rather than on the passage itself. The text itself should be the basis of the inference. This may seem challenging, but a little practice can go a long way in learning to make inferences. Observe the sentence below:

With only a few seconds left in the game, the coach was extremely nervous.

By reading the sentence above, one can infer that the game is really close, and that winning and losing are both possibilities. One cannot infer that the coach's team is winning or losing. Also, it cannot be inferred what sport is being played. Consider the addition of the following information:

> With only a few seconds left in the game, the coach was extremely nervous. The score was tied, and this was the state championship game. His team had the ball.

With the addition of more information, new inferences can be made. The reader now knows that the score is tied and that it is a state championship game. It is also a game with a ball. One can infer that the game is not hockey (a game with a puck and not a ball).

> With only a few seconds left in the game, the coach was extremely nervous. The score was tied, and this was the state championship game. His team had the ball. The star player dribbled the ball up the court and took a shot from the three-point line. Right as the ball left her hand, the buzzer went off, and the crowd held its breath to see if the ball would go through the hoop.

Now it can be inferred that the game is basketball. Words like "dribbling," "hoop," "three-point line," and "court" are used, providing evidence that the game is basketball. Since the star player is a woman ("her"), we can infer that this is a women's basketball team. We do not know who wins the game, but we can infer that there is a relatively large number of people watching, considering the author mentions that "the crowd held its breath." Inferences are based on evidence and implied information.

Conclusions

Drawing conclusions from the text is similar to drawing an inference in that one must use evidence from the text and not just one's own imagination. For reading comprehension tests, it is not uncommon for students to be asked to identify the most accurate conclusion from a given list of options. Typically, options with all-or-nothing verbiage can be eliminated. Therefore, selections with words such as *always* or *never* should be avoided. Keeping track of the pertinent points in the passage is important when forming a conclusion. Also, the conclusion should be directly supported by the text. The AFOQT will have five options to choose from for questions such as these, and it is important to select the one that is supported with evidence from the text.

Extrapolation

Extrapolation involves concluding the text with the assumption that things will continue in the current direction. Predicting the outcome of a text often requires using common sense. Consider the following example:

> She was so exhausted. After a long day of work, she turned off the light, slipped into bed, and closed her eyes.

The most logical prediction in the example above is that the woman fell asleep. Common sense and prior knowledge would indicate that exhausted people fall asleep when they turn off the lights, slip into bed, and close their eyes. This prediction may not be accurate, but it is the most likely next step. With more details, the prediction could change. Making an extrapolation here is pretty straightforward since everything is moving in one direction. The text indicates that sleep is the logical next step, but an additional sentence could change the prediction. The next sentence could read, "Just as she was falling asleep, everything in her room started to shake." Given the new details, the reader may predict that the woman will wake up with a shot of adrenaline. With new information comes an adjusted prediction.

Cause and Effect

Sometimes questions will require the test taker to identify the cause and effect in a given passage. The cause is an event that has a direct correlation to the outcome of something. The cause not only precedes the effect, but it is the reason for the effect. Words such as *because, since, due to, consequently, therefore, or as a result*, are all indicators of a cause and effect relationship. Example: The boy was crying (effect) because he fell down and scraped his knee (cause). Note that cause precedes effect in time, but not necessarily in the text. For example, here the crying is mentioned before the scraped knee, even though the scrape caused the crying.

Cause and effect relationships are relatively easy to identify, but sometimes they are not directly stated in the text. Consider the following: "The student had a test later in the day, and he was very stressed." The stress (effect) is obviously a result of the test coming up (cause); however, the text does not explicitly state that the stress was caused by the test coming up. This cause and effect relationship would be inferred by the reader. The cause and effect relationship is implied rather than stated outright.

Comparison and Contrast

Understanding the concept of comparison and contrast is vital to succeeding in any reading comprehension test. Readers will be expected to identify comparisons and/or contrasts in the text. Comparison involves connecting two or more objects or ideas by determining shared attributes. Comparison involves asking the question, "What do these have in common?" Contrast involves taking the same objects or ideas and determining what is not alike. Contrast involves asking the question, "What do these things *not* have in common?" Comparison and contrast are inversely related: one finds what is similar between two or more objects or ideas, and the other finds what is dissimilar.

Think about bananas and oranges. Comparison would identify that they are both fruits. Contrast would identify that bananas are yellow, whereas oranges are orange. Any time the given text relates what is similar about objects or ideas, comparison is happening. Any time the text relates what is dissimilar about the object or ideas, contrast is happening.

The Sequence

Determining the *sequence* in a given passage can be helpful in answering questions about the order of things. The order in which events happen in the text is considered the sequence. Words such as *initially, then, and last* indicate the sequence of events in the text. *First, second, and next* are words that indicate different stages or steps in the text. Questions that involve sequence require the reader to accurately identify the order in which things happen. To choose the correct option, the test taker must pay attention to details.

Sometimes the order of things is implied rather than definitively stated. Consider the following sentence:

He brushed his teeth, put on pajamas, and went to sleep.

In the above example, nowhere does it state the chronological order of events. The steps are not numbered. The reader is left to logically infer the accurate order of things. It would not make sense that the character would brush his teeth after going to sleep. Considering that the last logical action of the character would be to go to sleep, it is implied that the author wrote the sentence in chronological order.

Written passages are not always chronological. Keeping track of the sequence of events is important, especially when the sequence is implied or in reverse order. For practice, determine the sequence of events in the following paragraph:

Tomorrow she would give the most important presentation of her career. All the top executives in the company were going to be at the meeting, and she had to impress them with her research. In order to do well, it was important to get a good night's sleep. She was headed home from the lab and hoped to get to sleep within an hour.

The sequence is not chronological. The actual order of events is the following: driving home, going to sleep, executives coming to a meeting, and then giving the presentation/impressing them. If you are having difficulty determining sequence, one trick is to jot down a quick outline of the given passage.

Vocabulary in Context

Test questions may ask you to define a particular word in the context of the given text. Many times these words have various meanings or are uncommon. By analyzing the context of the word, it is possible to find clues as to what the word means. The context points toward the definition and sets parameters for what the word could mean. Consider the following and pay attention to the word in *italics*:

> The man was a *gourmand*. On his days off, he would eat and drink to no end. His food bill every month was through the roof.

Whether or not the word "gourmand" is familiar to you, the next couple of sentences suggest that it means "glutton." By looking at the context of the word, its definition is clear even if you've never seen the word before or know its meaning.

What happens if the definition is not so easy to discern given the context? Consider the following and pay attention to the word in *italics*:

> The army was expected to *capitulate*, but instead they attacked their enemy.

By using comparison and contrast, the reader can infer the meaning of the word *capitulate*. Even if the exact meaning is hard to discern, it can be inferred that it means something that is most likely the opposite of "attack." *Capitulate* means "surrender." When given an unknown word in the context of written text, look for clues. In this case you would look for an option that is an antonym of "attacked."

The Author's Position

Often, standardized reading comprehension tests will ask you to identify the author's position. The questions may start something like this: "Which of the following claims would the author most likely agree with?" To properly answer the question, you will have to discern the author's position. Sometimes this position is clearly stated, but other times it may be implied. As the reader, you will have to inspect the passage for clues as to what the author believes. If the author is using positive language when writing about the topic, it may be clear that the author is supportive of the subject. If the author is using negative language, the author's position would most likely be unsupportive of the subject. Pay attention to tone and connotation. Above all else, determine the author's position based on the overt statements in the text, but remember that many times you will also have to consider the tone and connotation of the text. Even academic articles that seem to have no tone or position are still written by an author with some type of bias. The author's position can be easy or hard to discern, depending on the passage of text.

Defining the Purpose

The purpose of the passage may closely relate to the author's position, but the two are separate. The position of the author is a belief and/or stance on the subject, but the purpose is why it was written. If the passage is primarily an intriguing story, then the purpose may be to entertain. If the passage is not emotional and more factual, the purpose may be to inform. If the passage contains much emotional language and intensity, most likely the purpose is to persuade. The purpose is the reason for the passage itself.

Additional Resources

If this portion of the test concerns you, check out some free resources at your local library. Libraries have many great resources for improving your reading comprehension skills, and a librarian can help you find the right materials.

Before You Start the Practice Questions

On the next page, you will begin a Reading Comprehension practice test. Set a timer for 38 minutes before you start this practice test. Giving yourself 38 minutes will give you an authentic feel for how long you have to finish this portion of the AFOQT. Like the official test, this practice test has exactly 25 items. Set your timer, turn the page, and begin.

Reading Comprehension Practice Test Questions
(38 Minutes)

1. Read the passage below and select the main idea:
    The story of the heavier-than-air flying machines really begins in the United States in the early 1890's. Octave Chanute, born in France and reared in America, was one of the first men to take a scientific approach to the problem of flying machines. A thorough scientist, he had followed the progress of all flight experiments the world over. He built gliders with one, two, and even five pairs of wings and tested all of them on the sand dunes of Lake Michigan.

    a. America is the birthplace of aviation.
    b. Octave Chanute was one of the first people to design flying machines.
    c. France had a massive influence on aviation.
    d. Planes were invented in Michigan.
    e. Planes are a result of worldwide innovative efforts.

2. Read the following paragraph and identify the most accurate meaning of the term "fuselage":
    The first step in correcting some of the faults of the early airplane came with the development of a body, or fuselage. The first fuselages were built of spruce frames covered with fabric and strengthened with wire. They were mounted between the wings and braced to them. The engine and propeller were housed in the front of the fuselage. Farther back, an enclosed compartment, or cockpit, was provided for the pilot. Thus, he was moved from his perch on the wing with the engine at his back into a safer and more comfortable location.

    a. The electronics of the plane
    b. The cockpit of the plane
    c. The body of the plane
    d. The wings of the plane
    e. The aircraft's protection

3. What can be inferred from the paragraph below?
    When war was declared in 1917, naval aviation consisted of 54 airplanes, 38 pilots, and 163 enlisted men. By rapid expansion, it reached the strength of more than 50,000 men and over 2,000 airplanes by the end of the war.

Some 17,000 men and 540 airplanes were sent abroad during the conflict. Extremely successful anti-submarine and patrol operations were carried out throughout the war, and our naval aviators served with great distinction.

   a. War destroyed innovation.
   b. Aviation grew exponentially during the war.
   c. Boats were not as important during the war.
   d. Submarines were not as often used during the war.
   e. Casualties were enormous during the war.

4. Read the passage below and identify the most accurate conclusion:
   Slow as she had been in starting, America picked up speed and finished World War I with a creditable record. American aviation discarded its swaddling clothes forever. At the time of the Armistice, American fliers had flown more than 3,500,000 miles in battle and dropped 275,000 pounds of explosives on the Germans. In plane-to-plane combat, U.S. military pilots showed courage and initiative unequaled by ally or foe.

   a. In the area of aviation, America was lagging behind during the war.
   b. The innovation of aeronautics was in its infancy stage.
   c. By the end of World War I, America had matured greatly in the area of aviation.
   d. America learned how to use airplanes better than the Germans did.
   e. Aviation was the key factor to winning World War I.

5. Read the paragraph below and identify the main idea:
   Fighter planes of World War I had an average wingspan of 28 feet and a ceiling of about 20,000 feet. They were powered by engines of 150 horsepower, with speeds ranging from 100 to 125 miles per hour. Their average weight was 1,500 pounds, they carried enough gasoline for a two hour flight, and they were armed with two .30-caliber machine guns. All of these planes had the habit of shedding parts under stress of battle, and more pilots were killed during the war because of defective equipment, lack of parachutes, and inexperience than as a result of enemy action.

   a. Fighter planes in World War I were state of the art.
   b. Aeronautics was a growing industry.
   c. Aviation was on the frontier of innovation.
   d. Fighter planes were more effective than other methods of combat.
   e. Planes in World War I were dangerous and imperfect.

6. What is the purpose of the passage below?

    It was not until the end of the war that Navy men began to consider the idea of building a surface vessel capable of carrying airplanes to sea. It was soon recognized that such a ship must be devoted exclusively to the carrying and handling of airplanes. It must literally be an aircraft carrier.

    a. To explain how ships were more effective than planes in war.
    b. To explain how ships and planes must work together for success.
    c. To describe the evolution of different kinds of naval ships.
    d. To explain how the idea of the aircraft carrier came to be.
    e. To describe how aircraft carriers were designed.

7. After reading the passage below, select the option that the author would most likely agree with:

    It was between April 6 and September 28, 1924, that the first flight around the world was made. Four Douglas Cruisers, each carrying two men, started the flight from Seattle, Washington. A world-wide organization was set up to service the planes as they circled the globe. Two of the planes completed the trip 175 days later. The total distance flown was 26,345 miles, and the total flying time was 363 hours, 7 minutes. A third plane was destroyed in a crash in Alaska early in the flight, and the fourth sank after a crash in the Atlantic on the last lap of the trip. The DWCs used in the flight were powered with 450-horsepower *Liberty* engines, and the average speed was about 72 miles per hour. This round-the-world flight was truly a daring operation.

    a. Flying in those days was exciting.
    b. Flying in those days was expensive.
    c. It was rare for people to fly.
    d. The first flight around the world was dangerous.
    e. The first flight around the world was innovative.

8. Read the following passage and identify the reason for which the government was about to abandon air mail:

    Difficulties had arisen in the air mail service by 1921. It had become apparent that air mail would not be valuable to the government unless it could be flown by night as well as by day. It had been standard practice for the mail to be flown only during daylight hours and to be carried by train at night. The government was about to abandon the air mail service when the pilots pointed out that all that was needed was a chain of airway beacons and lights for the landing fields and planes.

a. Mail was unpredictable from planes.
   b. Aircraft would often lose mail.
   c. Aircraft were too dependent on weather.
   d. Mail needed to be delivered at night.
   e. Mail was not needed every day.

9. Based on the paragraph below, select the claim about aviation that the author would most likely agree with:

   Just about the time the Ford tri-motors were proving themselves in tests, an important law was passed by Congress. It was the Kelly Air Commerce Act of 1925. It authorized the Post Office Department to contract with private firms to fly the air mail routes maintained by the Department of Commerce. This law was designed to encourage private capital to enter the aviation field, with the objective of carrying not only mail, but passengers, too. In February, 1926, officials of one of the newly-formed air transport firms proudly watched their first big air transport plane take off from the Detroit airport. The big plane was a Stout-designed, all-metal Ford, the first of a series of airliners that were destined to make aviation history.

   a. Private capital was discouraged.
   b. Private capital was used to advance aviation for mail and passengers.
   c. The Kelly Air Commerce Act was important for the environment.
   d. Ford was the global leader in aeronautics.
   e. The Post Office Department lost business to private firms.

10. What can be concluded from the following passage?

    The Wright Brothers' first airplane engine weighed 170 pounds and produced 12 horsepower. It used twenty-five percent of its energy propelling itself. With the introduction of the air-cooled, radial engine twenty years later, a pound and a half of engine had been made to produce one horsepower. Thus, the new 350-pound radial engine of 200 horsepower put all but a fraction of weight into load-carrying power.

    a. The Wright Brothers made superior planes.
    b. Heavier engines were advantageous.
    c. Most recent aircrafts were more efficient.
    d. Newer planes weighed less.
    e. Fuel efficiency was a primary concern for early aircrafts.

11. Based on the following paragraph, determine the most accurate characteristic of the Wright Brothers:

    After a year of exhaustive study and experiments with models in their wind tunnel, the Wright Brothers were ready to experiment with a man-carrying glider. With the thoroughness that was typical of every move of the Wrights, the brothers asked the government to let them have information on meteorological conditions all over the country. By studying the weather charts, they were able to find a locality where there was a continual flow of wind. This would be nature's wind tunnel, where they could test their glider day after day.

    a. The Wright Brothers were very thorough in their research.
    b. The Wright Brothers were risk-takers.
    c. The Wright Brothers were ahead of their class.
    d. The Wright Brothers needed more wind.
    e. The Wright Brothers were perfectionists.

12. What can be inferred from the passage below:

    The best fighter planes used by the Germans in World War I were not of German design. They were designed and built under the supervision of a young man from Holland. Tony Fokker had offered his airplane designs to his native Holland. They were refused. In turn, Fokker tried to interest the British, French, and Belgians in his airplanes, but none of them took him seriously. Just before World War I, the Germans "tied up" Fokker with a contract that practically kept him their prisoner until the war was over.

    a. The Germans did not know how to make planes, so they needed Tony Fokker.
    b. Holland produced Germany's planes during World War I.
    c. Germany only had Fokker's designs because other countries had turned them down.
    d. Germany made Tony Fokker a prisoner of war.
    e. World War I would have never started if Holland would have just accepted Fokker's plane designs.

13. After reading the passage below, select the most likely reason why pilots and operation men were not keen about carrying passengers:

    During 1927 and 1928, the map of the United States showed a continually increasing number of lines marked "Air Mail Route." In 1926, the sixteen companies holding airmail contracts flew about 1,700,000 air miles. Much of this mileage was flown in single-engine, open-cockpit airplanes. Mail was the principal source of revenue. The few passengers who first braved the rigors of early air transport either rode on mail sacks or in small, cramped cockpits. Pilots and operation men alike frankly admitted they were not keen about carrying passengers.

a. Passengers did not pay their fair share.
b. There was not very much room in airplanes.
c. Pilots and operation men didn't want the liability of carrying passengers.
d. They could make more money by hauling more mail if passengers did not ride along.
e. Pilots and operation men were distracted by the number of passengers onboard.

14. According to the following passage, what can be concluded about air travel in 1928?

In 1928, an air traveler making an extensive trip would be likely to fly in seven or eight different types of planes. He might step into a Fokker tri-motor, change to a single-engine Boeing, ride for some distance in a Ford tri-motor or a Whirlwind-powered *Travel Air*, and finish his trip in a Curtiss *Carrier Pigeon*. The planes usually flew low, at between one and two thousand feet. Here the air was usually rough, and a good percentage of air travelers were troubled with airsickness. The planes landed every few hundred miles to refuel. They were noisy and heated only by exhaust gases from the engines, which usually furnished more sickly fumes than heat. Little food, if any, was served, and a coast-to-coast journey took thirty-three hours.

a. Flying was adventurous.
b. Flying was expensive.
c. Flying was dangerous.
d. Flying was luxurious.
e. Flying was difficult.

15. In the following paragraph, the word "industrious" most nearly means:

Although Donald Douglas had achieved a great deal of international fame as a result of his round-the-world flight and was highly respected in military circles, few other people knew him. A quiet, industrious young man, he put all his earnings back into his business and continued to work on his dream of big, roomy, smooth-flying airliners. He visualized air transport flying from coast to coast and from country to country in a great network of airlines that would link the whole world.

a. Skilled
b. Expansion
c. Visionary
d. Hard-working
e. Specialized

16. According to the following passage, what was the purpose of a gyroscope in aviation?

> The application of the gyroscope to aircraft instruments was a great step in the advancement of flying. First experimented with by Lawrence Sperry in the early days of the airplane, the constant action of the gyroscope was used to register the changes of attitude of an aircraft in flight. It was first used in the turn and bank indicator, then in the *gyro horizon and directional gyro*. Power-driven gyros constantly whirled in the direction in which they were set. They were attached to dials on the instrument panel and the plane itself.

a. To determine speed of aircraft
b. To measure air pressure
c. To measure the power of an engine
d. To determine direction
e. To measure the change in attitude

17. Select the option that the author of the following paragraph would most likely agree with:

> In building the two-engine, all-metal B-9, Boeing engineers learned how to build another plane with a more peaceful purpose. This ship was the famous Boeing 247-D commercial transport plane. The 247-D was an all-metal, low-wing monoplane, powered with two 550-horsepower Pratt & Whitney Wasp radial engines. It had a top speed of 200 miles per hour and a cruising speed of 180 miles per hour. It was America's first three-mile-a-minute air transport plane.

a. The Boeing 247-D was designed for war.
b. The Boeing 247-D was state of the art.
c. Boeing engineers wanted world peace.
d. The Boeing 247-D was the fastest plane in the world.
e. The B-9 was a faulty plane.

18. What can be concluded from the following passage?

   The air mail revenue was the lifeblood of the air transport operators, and the cancellation of mail contracts suddenly darkened their future. An attempt to put the transportation of air mail into the hands of the United States Army resulted in a tragic failure. This was mainly due to the unfamiliarity of Army pilots with air mail routes and a lack of proper equipment. In June, 1934, air mail was turned back over to the airlines.

   a. Army pilots were the lifeblood of the air mail industry.
   b. The government is best at delivering air mail.
   c. Airlines would rather transport passengers than air mail.
   d. Private airlines almost went bankrupt.
   e. Private airlines were best at delivering air mail.

19. Based on the passage below, what can be inferred about airplanes before World War I?

   In the very early days of aviation, before the start of World War I, most of the airplanes, with the exception of a few military ships, were sold to private owners. Those buyers were either barnstormers or wealthy sportsmen. Some advertising in national magazines even tried to create sales for private planes. This activity ceased with the beginning of the war in 1914, and owners turned their planes over to the government for training purposes.

   a. Aviation was not a common aspect of war.
   b. The government could not afford planes.
   c. Planes were not reliable.
   d. Private planes were outlawed.
   e. Airplanes were used for hunting.

20. According to the following paragraph, the Boeing 307 was:

   The Boeing Company, working in cooperation with Tomlinson, Transcontinental and Western Airways, and Pan American, developed the Boeing 307. The 307 was a big, all-metal, low-wing monoplane with a pressurized, high-altitude cabin, which made flight at altitudes up to 20,000 feet possible. This was accomplished in a manner similar to that used in supercharging the engines. Engine-driven superchargers pumped air into the cabin-ventilating system, and the atmosphere in the plane was kept at normal low-level pressure regardless of how high the plane flew. The Boeing 307 *Stratoliner* was put into service by TWA and Pan American

Airways in 1940 and marked a tremendous step forward in the speed and comfort of modern air travel.

a. Uncomfortable
b. Unrivaled
c. Dangerous
d. Expensive
e. Outdated

21. The purpose of the passage below is to:
    Experiments with the use of aerial torpedoes brought about the development of the Douglas TBD-1 torpedo plane. Though not as fast as a fighter, the three-place TBD-1 *Devastator* carried a deadly torpedo load. The Douglas SBD *Dauntless* was designed for dive-bombing and was the first low-wing monoplane to be used as the standard dive-bomber on U.S. carriers.

    a. Explain how torpedoes were manufactured.
    b. Describe the development of dive-bombers and torpedo planes.
    c. Describe the dangers of torpedo experiments.
    d. Explain the difficulties of fighter plane innovation.
    e. Describe the importance of U.S. aircraft carriers.

22. What can be concluded from the paragraph below?
    It was not until the outbreak of World War II that most Americans came to realize the value of the airplane in modern conflict. As the fighting grew to global proportions, Americans began to appreciate the farsightedness of our Air Corps leaders, particularly in developing the long-range bomber.

    a. Aeronautical advancements won World War II.
    b. Traveling by boat became rare.
    c. Airplanes become crucial to military success.
    d. Americans appreciated being able to fly.
    e. Airplanes were in high demand.

23. What does the term "farsighted" mean in the passage below?
    From small raids by a dozen *Fortresses,* the number of bombers grew until the raids became huge aerial invasions involving hundreds of bombers and thousands of airmen. That the path for invasion was cleared and victory brought nearer was due in no small measure to our big bombers and the

farsighted American airmen who had brought them into being against almost insurmountable obstacles.

a. Being able to see a long distance
b. Unable to see things up close
c. Ambitious
d. Imagining the future
e. Tenacious

24. According to the following passage, how did the use of airplanes in combat evolve?

    Although the airplane in World War I had been used mainly as an observation and a plane-to-plane combat weapon, wise American airmen, such as General "Billy" Mitchell, visualized the craft as a means of destroying the enemy's ability to fight. As the result of this thinking, the doctrine of air power was established.

    a. Fighter planes were in lower demand.
    b. Using planes for observation was outdated.
    c. Planes were only used for bombing the enemy.
    d. Planes were used for new purposes.
    e. In those days, airplanes were used to generate electricity.

25. The Northrop P-61 Black Widow night-fighter was:

    The year 1944 saw a twelve-and-one-half ton fighter go into action on the war fronts. This plane, the Northrop P-61 *Black Widow* night-fighter, was one of the most powerful airplanes yet to go into action. Powered with two 2,000-horsepower engines, the P-61 flew at 400 miles per hour. Equipped with radar and powerful guns, it could search out an enemy plane at night and destroy it.

    a. Only used at night
    b. One of the most powerful planes of its time
    c. Useless in times of peace
    d. Impossible to locate
    e. The primary military advantage of the U.S.

Answer Guide to Reading Comprehension Practice Test

1. **B.** This passage is about scientist Octave Chanute experimenting with heavier-than-air flying machines. It does not say that aviation was birthed in America. This passage is not about France's influence on aviation. This passage does not say that planes were invented in Michigan. It may be true that planes are a result of worldwide innovative efforts, but that is not what this passage is about.
2. **C.** The fuselage is described as the body of the plane. It may include the cockpit, but it is not synonymous with "cockpit." It may protect the plane to a degree, but the passage describes it as the plane's body.
3. **B.** This passage describes the expansion of aviation during war. None of the other options were mentioned in the passage. Casualties were enormous during the war, but that truth was not stated anywhere in the paragraph.
4. **C.** The paragraph describes how America went from a slow start to domination in the realm of aviation. Option D wasn't stated in the paragraph, though it may be true. Option E was not stated in the paragraph, although it may be true. This was a passage about America maturing and "discarding its swaddling clothes (what a newborn baby wears) forever."
5. **E.** Pilots were killed more often by defective equipment, lack of parachutes, and inexperience than they were by enemy action. This describes danger and imperfection. The connotation of the passage is not to describe how state of the art the planes were—quite the contrary. Aeronautics was a growing industry, but nothing in the passage indicates that truth. Option C was not mentioned in the passage. Option D was not stated in the passage.
6. **D.** The purpose of this passage is to give some background on how the idea of the aircraft carrier came to be. Option A was not mentioned in the passage. Option B could be inferred by the passage, but it was not the primary purpose of the paragraph. Option C is not the correct answer because this passage only talks about the aircraft carrier; the passage is not about different kinds of naval ships or their evolution. The design of the aircraft carrier is not mentioned in the passage, only a general idea of the concept.
7. **D.** The paragraph talks about how two of the four planes trying to fly around the world crashed. It mentions how it was a daring operation. As exciting as it may have been, nowhere in the passage is this emotion described. The price of the tickets was not mentioned. The rarity of people flying was not discussed. The level of innovation it took to fly around the world was not a focal point.
8. **D.** The paragraph clearly states that the government needed mail to be delivered at night. None of the other options were mentioned in the passage. Some of the other options may be true but were not expressed in the paragraph.

9. **B.** The passage describes how the Kelly Air Commerce Act of 1925 was "designed to encourage private capital to enter the aviation field, with the objective of carrying not only mail, but passengers, too." None of the other options were explicitly stated in the passage.

10. **C.** This passage talks about how more recent engines could produce more horsepower per pound of engine. They were more efficient. None of the other options are discussed in the passage. Option D could be inferred by the passage, but it is not stated. The passage describes how the newer planes had a 350-pound engine and the Wright Brothers' first airplane had a 170-pound engine. Planes overall were getting bigger, not lighter, but they had greater capacities.

11. **A.** The Wright Brothers were very thorough in their research, as described in the passage. The other options describe characteristics that may be true but were not outlined in the passage. Option D was described in the passage, but is not a characteristic of the brothers themselves; it was a need they had for their research.

12. **C.** The passage clearly explains how Fokker contracted his designs to Germany after being denied by other countries. It can be inferred that Germany would not have had the designs if other countries had not turned him down. None of the other options were mentioned in the passage, nor could they clearly be inferred by the passage.

13. **B.** Out of all the options, the only thing stated in the passage was that the passengers rode in small, cramped cockpits. Based on the passage, this could be the only reason. The other options may be true but could only be selected by conjecture.

14. **E.** Airsickness, noise, little food, and long journeys were all factors that contributed to the difficulty of air travel. It may have been adventurous, expensive, dangerous, and at times luxurious, but none of this was mentioned in the text.

15. **D.** "Industrious" is an adjective that means "diligent" or "hard-working." He may have been a skilled, visionary, or specialized person, but "industrious" in this passage means "hard-working," as he "put all his earnings back into his business and continued to work on his dream."

16. **E.** The passage indicates that gyroscopes are used to measure the change of attitude during flight. The passage does not indicate any of the other options as having been the purpose of a gyroscope in aviation.

17. **B.** The implication of this passage is that the Boeing 247-D was state of the art at that point in history. None of the other options can be found in the passage. We do not know that it was the fastest plane in the world; we only know it was the fastest "air transport plane." Since it is categorized like this, there were most likely other planes that were faster but not "air transport plane[s]."

18. **E.** The government handed the air mail routes back to private airlines after failing to maintain them. No other options were mentioned in the passage. The passage

mentions that the future for air transport operators looked dark, but it makes no mention of bankruptcy.

19. **A.** Most planes were for private purposes but started becoming more useful during World War I. None of the other options are mentioned in the passage. Sportsmen may have purchased planes, but that does not necessarily mean they used the planes for hunting. Most likely, planes were used recreationally, based on the passage.

20. **B.** The main point of this passage is to explain how awesome the Boeing 307 was. No other option would make sense after reading the passage.

21. **B.** No mention is made of torpedoes being manufactured or the dangers of experimenting with them. The importance of U.S. aircraft carriers and the difficulties of the fighter plane innovation are not discussed. This passage describes the development of dive-bombers and torpedo planes.

22. **C.** According to the paragraph, military success was related to aviation. Some of the options may be true, but they are not discussed in the passage. Aeronautical advancements may have helped, but they did not necessarily win the war. No mention of boats is in the passage. Americans may have appreciated the ability to fly, but the passage makes no mention of it. Airplanes probably were in high demand, but the passage does not say so. There may or may not have been enough supply, but the paragraph does not indicate anything of the sort.

23. **D.** The term "farsighted" can mean several of the listed options, but in the context of this passage it means "imagining the future." It talks about how airmen "brought them into being." This speaks of manifesting an idea or concept. In that line we find the clue that "farsighted" means something related to ideas or foresight, not literal eyesight. Options A and B are not the answer because they relate to literal eyesight, which is not what this passage is referring to. Options C and E may be other characteristics of the visionary airmen, but they are not synonymous with "farsighted."

24. **D.** The passage mentions that planes were used for more than plane-to-plane combat and observation. None of the other options are mentioned in the passage. Option E is misleading because the term "air power" is not referring to electricity. Air power is a doctrine. Planes were used for bombing, but not only bombing. The passage makes no mention that planes used for observation were outdated. Also, nothing is mentioned about fighter planes being in lower demand.

25. **B.** The passage states that the P-61 was one of the most powerful airplanes yet to go into action. It was used at night, but the passage does not indicate that it was used only at night. The other options may be conjecture, but they are not indicated in the passage.

SITUATIONAL JUDGMENT

The Situational Judgment portion of the AFOQT tests interpersonal skills. This section contains different scenarios and various possible responses. Both the most effective and least effective responses must be selected. These scenarios are typical situations that officers encounter when working in their roles. Analyzing each scenario and selecting the most and least effective actions will help determine the test taker's ability to handle professional relationships.

Self-sufficient decision-making and good judgment are two crucial traits of a successful officer. No specific material can be reviewed to prepare for this part of the test, but general leadership qualities such as integrity, professionalism, resourcefulness, and innovation should be kept in mind when answering questions.

The best response to each situation can be subjective at times. For this reason, the Air Force has taken a census of experienced Air Force officers to help measure the effectiveness of each response to different situations.

When selecting the best course of action for each given situation, try to choose the most logical response. Rely on logic more than emotion. Select the action that will result in resolution. Actions should not lead to new issues in the future. Problems should be solved in the quickest and easiest way possible.

Before You Start the Practice Questions

On the next page, you will begin a Situational Judgment practice test. Set a timer for 35 minutes before you start this practice test. Giving yourself 35 minutes will give you an authentic feel for how long you have to finish this portion of the AFOQT. Like the official test, this practice test has exactly 25 situations with 50 selections (2 for each situation). Set your timer, turn the page, and begin.

Situational Judgment Practice Test Questions

(35 Minutes)

Situation 1

Each airman in your unit rotates through a list of assignments, some of which are more pleasant than others. You notice that your superior never assigns the worst task to his friend in the unit and often assigns it to his least favorite airman.

Possible responses:

- a. Confront your superior about his favoritism.
- b. It's not really your problem, so simply ignore it.
- c. Offer to help the airman handle the task.
- d. Take the problem to your superior's commanding officer.
- e. Favor the friend, so you can avoid the task yourself.

1. Choose the MOST EFFECTIVE response to the situation.
2. Choose the LEAST EFFECTIVE response to the situation.

Situation 2

At a night out with your team, one airman spills a secret about your commanding officer. It seems on several occasions the individual saw the officer in question drink whiskey on duty from a bottle stashed in his desk.

Possible responses:

    a. Confront the officer about the allegations.
    b. It's hearsay at this point, so simply ignore it.
    c. After the officer's shift, search his desk for alcohol as proof.
    d. Take the problem to your commanding officer's superior.
    e. Tell the airman to stop spreading rumors or he will face discipline.

3. Choose the MOST EFFECTIVE response to the situation.
4. Choose the LEAST EFFECTIVE response to the situation.

Situation 3

An airman who has performed brilliantly in action is up for a well-deserved promotion. He needs to get a certain grade on a test to be considered, but he has a bad day and comes up short by 20 points. You know he's more than qualified for the promotion, and there won't be another opportunity for promotion for a while.

Possible responses:

    a. Give him 100% on the test, so he gets promoted.
    b. Explain the situation to your commanding officer and ask for advice.
    c. Stick to the rules and give the airman a failing grade.
    d. Let the airman take the test again to get a better grade.
    e. Take no action and let things work out as they will.

5. Choose the MOST EFFECTIVE response to the situation.
6. Choose the LEAST EFFECTIVE response to the situation.

Situation 4

A new airman has joined your squadron, and he has a bad attitude. He starts fights and slacks on his duties. Morale is at an all-time low.

Possible responses:

a. The squadron must learn to self-regulate, so let them work it out amongst themselves.
b. Explain the situation to your commanding officer.
c. Transfer the troublemaker quickly to another squadron.
d. Meet with the new airman to enforce a degree of disciplinary action and explain the consequences if he doesn't change his attitude and efforts.
e. Schedule a squadron-wide meeting to talk about attitude adjustments.

7. Choose the MOST EFFECTIVE response to the situation.
8. Choose the LEAST EFFECTIVE response to the situation.

Situation 5

Several officers need your signature on promotion paperwork for a few airmen. You sign the paperwork and send it on to your superior. Afterward, you discover that one individual being promoted doesn't meet the qualifications.

Possible responses:

a. Go immediately to your superior with the information.
b. Talk to the officer promoting the airman to find out why he's being promoted.
c. Figure there must be a good reason for the promotion and do nothing.
d. Request all paperwork back and check everyone's qualifications.
e. Schedule an officers' meeting to talk about how best to handle the paperwork.

9. Choose the MOST EFFECTIVE response to the situation.
10. Choose the LEAST EFFECTIVE response to the situation.

Situation 6

As an airman in your squadron gets close to his transfer date, he starts slacking off on his responsibilities. You've mentioned it twice, and he apologized, but nothing changes. The other airmen are getting irritated at picking up the slack. The transfer is scheduled for next week.

Possible responses:

a. Ignore the situation because the airman will be gone in a week.
b. Discipline the airman and put a report in his file.
c. Ask your commanding officer for his recommendation.
d. Threaten to rescind his transfer if he doesn't improve his habits.
e. Let the squadron know you expect them to handle the situation.

11. Choose the MOST EFFECTIVE response to the situation.
12. Choose the LEAST EFFECTIVE response to the situation.

Situation 7

Your entire squadron spends a week of training in new technology, which they must understand to safely do their jobs. One airman skips a day of training and spends the day with his kids at the zoo. You don't find out until after the training is complete and all airmen have received completion certificates.

Possible responses:

a. Make the airman retake the training program.
b. Ask your squadron to bring him up to speed on what he missed.
c. Talk to your commanding officer about charging the airman with AWOL.
d. Threaten to deny all future leave until he transfers to another squadron.
e. Do nothing because he already has his completion certificate.

13. Choose the MOST EFFECTIVE response to the situation.
14. Choose the LEAST EFFECTIVE response to the situation.

Situation 8

A new officer joins your unit, and the commanding officer requests that each officer help the new one out with whatever she needs. The new officer takes advantage of this directive by having the other officers do much of her work. When you bring it up to your commanding officer, he disagrees with your view and says she's simply asking for help until she gets more comfortable with her new role.

Possible responses:

a. Complete a report about the issue and send it to your commanding officer's superior.
b. Submit to your commanding officer's directive, but if the problem continues, meet with the new officer to express your concerns.
c. Gather the support of your peers and disregard your commanding officer's decision.
d. Confront the new officer and call her out for her laziness.
e. Avoid taking any action and keep your head down.

15. Choose the MOST EFFECTIVE response to the situation.
16. Choose the LEAST EFFECTIVE response to the situation.

Situation 9

Your unit's network is scheduled to be upgraded during the middle of the work day, holding everyone up in getting their tasks done. IT estimates it will take at least 4 hours. You think this inconvenient scheduling is in retaliation for your not supporting the head of the IT department during a recent request for additional funding.

Possible responses:

a. Threaten to take this to the highest level if the IT department doesn't reschedule during off-peak hours.
b. Tell your unit to put work on hold while the network is being updated.
c. Schedule a meeting with the head of IT to talk through the issue and schedule a better time for all involved.
d. Send an email to the head of IT, copying his superior, to accuse him of retaliation.
e. Inform your unit that they will still be expected to finish their tasks for the day, even if it means working overtime.

17. Choose the MOST EFFECTIVE response to the situation.
18. Choose the LEAST EFFECTIVE response to the situation.

Situation 10

Your commanding officer's superior comes to you with a problem. The superior has received complaints of your commanding officer treating female airmen unfairly, sometimes even aggressively, but you haven't witnessed this. He wants proof of these accusations and asks you to observe your commanding officer and report back to him as soon as possible.

Possible responses:

a. Politely refuse to "spy" on your commanding officer, telling his superior you've noticed nothing of the sort.
b. Observe your commanding officer and document any unfair treatment of women in a written report.
c. Explain to your commanding officer's superior that he could certainly improve but that the accusations are grossly overinflated.
d. Only report positive information to your commanding officer's superior.
e. Use this opportunity to lambast your commanding officer.

19. Choose the MOST EFFECTIVE response to the situation.
20. Choose the LEAST EFFECTIVE response to the situation.

Situation 11

After being promoted a year ago, you've performed at an exceptional level and your commanding officer is highly complimentary. He recently sent you a plan he devised to boost a unit's performance while on duty and told you to implement it. But the plan has caused significant problems that are beyond your control to rectify and unit performance is slipping instead. Your commanding officer states the problems are a result of your improper management.

Possible responses:

a. Tell your commanding officer that the plan is flawed, leaving you with no chance to succeed.
b. Talk to your fellow officers about how they're handling the situation, and see if they have suggestions for improvement.
c. Blame the results on your subordinates and take them to task.
d. Ask your commanding officer for suggestions on how to handle the particular problems you're encountering.
e. Request a transfer, so you don't have to work with this commanding officer anymore.

21. Choose the MOST EFFECTIVE response to the situation.
22. Choose the LEAST EFFECTIVE response to the situation.

Situation 12

During a routine audit, you notice discrepancies in the computer program that tracks your inventory compared to the Bill of Sale for certain supplies ordered through your system. When you log in to the ordering system, you see someone has changed figures to look like your unit received fewer supplies than what's being paid for. Only your peer officers have access to the ordering system.

Possible responses:

a. Schedule a meeting with your peer officers to outline what you've found and ask if anyone knows anything about it.
b. Take the problem to your superior and ask him to help find the culprit.
c. Ask your IT technician to do some digging and see which computer the changes were made from.
d. Ignore the problem because you don't know who's doing it, and it's really not that big of a deal.
e. Request a change so only a single officer has authorization to access the ordering system, making it more secure.

23. Choose the MOST EFFECTIVE response to the situation.
24. Choose the LEAST EFFECTIVE response to the situation.

Situation 13

You completed a report for your commanding officer that contains classified information about a new mission. After your shift, when you get back to your living quarters, you realize you forgot to pick up the printed copy of the classified report from the office printer.

Possible responses:

a. Show up early the next morning to get the report before others arrive.
b. Call the base and see if anyone is still in the office and ask them to put it in your desk drawer.
c. Immediately head back to the base to secure the classified report.
d. Send a subordinate to the base to pick up the report and bring it to your living quarters.
e. The likelihood is slim that anyone will see it, so get it in the morning.

25. Choose the MOST EFFECTIVE response to the situation.
26. Choose the LEAST EFFECTIVE response to the situation.

Situation 14

A fellow officer comes to you for help. She's been working overtime for weeks to get some special reports completed by their due date. As a result, the rest of her workload is piling up on her desk. She wants to know if you can help her.

Possible responses:

 a. Explain that you have your own work to do, so you can't help her.
 b. Go to your superior to explain the situation and ask him to reassess her workload.
 c. Tell your fellow officer you can have one of your subordinates pick up the slack.
 d. Offer to help her after you've finished your work.
 e. Tell her to try the other officers because they have more time than you do.

 27. Choose the MOST EFFECTIVE response to the situation.
 28. Choose the LEAST EFFECTIVE response to the situation.

Situation 15

You've been transferred to a new assignment and are still in the transition phase. One of your peers comes to you with a report he created and asks you to look it over. He's presenting it to his superiors the next day. You point out a few of his ideas aren't feasible with current staffing and budgets. He tells you you're too new to the position to know what's feasible and refuses to listen to your explanation.

Possible responses:

 a. Let it go. You tried to help him, and he refused your help.
 b. Keep trying to persuade him to listen to you.
 c. Go to your senior officer with the situation and ask her what you should do.
 d. Do nothing because he deserves to fail.
 e. After the presentation, make it clear to your senior officer that you had tried to tell your peer he was wrong.

 29. Choose the MOST EFFECTIVE response to the situation.
 30. Choose the LEAST EFFECTIVE response to the situation.

Situation 16

One of your friends in another unit has been pressuring you to transfer to her unit. You've been in your current unit for 36 months, and while you've been happy where you are, you're starting to think it might be nice to have a change. Your friend is pushing hard for your transfer and rumors abound that you're actively seeking a transfer to her unit.

Possible responses:

 a. Ignore the rumors and they'll stop eventually.
 b. Continue researching the transfer while vehemently denying the rumors.
 c. Go to your senior officer with the situation and ask her what you should do.
 d. Schedule a meeting with your peers to discuss the pros and cons of a transfer.
 e. Admit that the rumors are true and actively seek a transfer.

 31. Choose the MOST EFFECTIVE response to the situation.
 32. Choose the LEAST EFFECTIVE response to the situation.

Situation 17

Your superior's aide is highly successful in her role, and you find out she's up for promotion to a new unit. But your superior doesn't know that she's been vocally critical of the unit's senior officer she'll be transferred under. In particular, she seems to reserve her most abrasive criticism for this senior officer's authority and is disrespectful to several of his subordinates.

Possible responses:

 a. Send your superior an email describing the aide's behavior and attitude.
 b. Highly recommend the promotion to your superior because she has to learn to work with others, even those she's critical of.
 c. Do not try to affect your superior's decision, but give your honest opinion of the aide's behavior.
 d. Schedule a meeting with your peers to discuss how you can sabotage the aide's promotion.
 e. Meet with your superior to explain why he should consider changing her promotion to another unit.

 33. Choose the MOST EFFECTIVE response to the situation.
 34. Choose the LEAST EFFECTIVE response to the situation.

Situation 18

Two of your peer officers are competing for the same promotion. Each is actively working to undermine the other's performance and reputation. It is counterproductive for the squadron and is affecting morale. Your commanding officer, unaware of the attempt at undermining, is undecided and asks for your input on who would best serve in the new role.

Possible responses:

- a. Decide who you like most and recommend them.
- b. Tell your commanding officer you refuse to get involved in their dispute.
- c. Go to your commanding officer, explain the situation, and ask how to proceed.
- d. Schedule a meeting with all your peers to talk about how best to get along with each other.
- e. Determine who can help your career the most, and choose them.

35. Choose the MOST EFFECTIVE response to the situation.
36. Choose the LEAST EFFECTIVE response to the situation.

Situation 19

You've recently taken command of a new squadron. On your first week in command, one airman is blatantly disrespectful and insubordinate. You decide to write him up, until he mentions that your superior is his uncle, who won't be happy.

Possible responses:

- a. You don't know your superior well enough to know how to handle this situation, so you do nothing.
- b. Try to work it out with the airman, because you need to be on his good side.
- c. Go to your superior's peer, explain the situation, and ask how to proceed.
- d. Tell the airman he has one strike against him, but you'll let it slide this time. Then put in for a transfer to another squadron.
- e. Write the airman up for insubordination and send it on to your superior.

37. Choose the MOST EFFECTIVE response to the situation.
38. Choose the LEAST EFFECTIVE response to the situation.

Situation 20

While off duty in town one night, several airmen in another unit get in a fight with civilians in a bar. The bar is severely damaged and several people are injured, but you've heard the airmen were defending themselves. A reporter for the local news station corners you and asks what the base is going to do about the rampant destruction of property and the hostility of its airmen.

Possible responses:

a. Explain it's not your place to render an opinion and give her the name of the unit's commanding officer.
b. Lead the reporter to your superior to ask for her opinion.
c. Even though these aren't your subordinates and you might not have all the details, give your honest and candid opinion.
d. Give a positive spin to the situation and provide the reporter with a sunny response.
e. Explain that these airmen are not under your command and it would be irresponsible to speak without having all the facts.

39. Choose the MOST EFFECTIVE response to the situation.
40. Choose the LEAST EFFECTIVE response to the situation.

Situation 21

You're halfway through a 6-week assignment with a fellow officer when he suddenly takes sick leave. He's going to be out at least two weeks, possibly three. There's no way you can finish the assignment by yourself.

Possible responses:

a. Meet with your supervisor to explain the situation and ask for an extension from the time your fellow officer gets back from sick leave.
b. Work as much overtime as you can to finish the assignment on time.
c. Command two of your subordinates to take the officer's place in the assignment.
d. Send your superior an email explaining the situation and asking for his advice.
e. Do what you can and hope your fellow officer gets back in time to help you finish the assignment.

41. Choose the MOST EFFECTIVE response to the situation.
42. Choose the LEAST EFFECTIVE response to the situation.

Situation 22

Your base's computer network is down, and your IT crew is working hard to bring it back up as quickly as possible. They've ordered new equipment to fix the problem, but it won't arrive until tomorrow. Your commanding officer must have a specific report by the end of today, but it's on the network, so you can't access it.

Possible responses:

a. Find a laptop and retype the entire report by the end of the day to meet your commanding officer's deadline.
b. Call your commanding officer and ask for an extension until the end of the day tomorrow, due to circumstances beyond your control.
c. Try to get in extra early the next morning to finish the report and have it on your superior's desk before he gets in.
d. Stick your head in your superior's office and explain you'll get him the report as soon as the network comes back up, without telling him it's likely to be tomorrow.
e. Ignore the situation because it's out of your control.

43. Choose the MOST EFFECTIVE response to the situation.
44. Choose the LEAST EFFECTIVE response to the situation.

Situation 23

Your transfer to a new unit is in a few days. The commanding officer sends you an email explaining that he can take a few hours to explain your duties to you. After that, it's up to you to make sure you're trained properly because he needs you to hit the ground running. Several things are due in the first week at your new unit.

Possible responses:

a. Find an internet video showing how to do what's being required of you.
b. Respond to the email asking for a peer or other resource to help you learn everything you need to know.
c. Decide you don't want to work with this superior and request to terminate your transfer.
d. Schedule a meeting with your new peers to ask if this is the way the commanding officer always handles things.
e. Do nothing, hoping that it will all work out.

45. Choose the MOST EFFECTIVE response to the situation.
46. Choose the LEAST EFFECTIVE response to the situation.

Situation 24

Your superior assigns you to man the gate where civilians come through. You haven't received the list of expected civilians, so you have no way of telling who is authorized to be on base and who is not. The civilians are becoming angry and insist you fix this problem immediately.

Possible responses:

a. Contact the individuals responsible for supplying the list and ask for the list as soon as possible.
b. Get paper and pen to capture who entered the gate and hope it matches the list of expected civilians.
c. Call an airman to take your place at the gate because you're suddenly not feeling well.
d. Try to convince the civilians to go get some coffee and come back to the base in a few hours.
e. Call your supervisor, explain the situation, and ask what you should do.

47. Choose the MOST EFFECTIVE response to the situation.
48. Choose the LEAST EFFECTIVE response to the situation.

Situation 25

Your squadon has a new commanding officer, and the rumor is that he's dating an airman in your unit. You haven't observed anything overt, but you notice the airman and commanding officer appear to know each other well. You have your suspicions but no solid proof.

Possible responses:

a. Call the airman out during a squadron-wide meeting and berate her.
b. Schedule a one-on-one meeting with the new commanding officer to let him know of the rumors.
c. Go over your new commanding officer's head with a report detailing your suspicions.
d. Until you have proof of an inappropriate relationship, ignore the rumors and keep your head down.
e. Meet separately with other airmen and officers to gather evidence of wrong-doing.

49. Choose the MOST EFFECTIVE response to the situation.
50. Choose the LEAST EFFECTIVE response to the situation.

Answer Guide to Situational Judgment Practice Test

1. **A (Most effective)** Talking directly with your superior about what you've noticed would be the best first step in bringing attention to the problem. Maybe your superior doesn't realize how frequently it happens.
2. **B (Least effective)** If you do nothing, the situation will continue, negatively affecting morale in the entire unit.
3. **A (Most effective)** Take the rumor directly to your commanding officer. He should have the chance to either prove the rumor false or take steps to change the situation.
4. **B (Least effective)** Doing nothing won't stop the rumors or the behavior, if the rumor is true.
5. **C (Most effective)** Part of being a leader is enforcing the rules even when it's difficult. Your integrity is at stake; stick to the rules.
6. **A (Least effective)** Compromising your integrity as an officer by falsifying test scores would affect your reputation as a leader.
7. **D (Most effective)** Sometimes subordinates need a wake-up call. You want your subordinates to succeed, but explaining disciplinary consequences helps them better understand their role and actions.
8. **B (Least effective)** As a leader, you must be able to manage behavior problems on your own. Taking this situation to your superior undermines your reputation as a competent leader.
9. **A (Most effective)** You're the one who made the mistake. It's important to rectify it as soon as possible by going to your superior with the information and requesting the information back so you can make corrections.
10. **C (Least effective)** Doing nothing undermines your credibility and could negatively impact your career.
11. **B (Most effective)** A large part of being a leader is handling behavior problems effectively. Since you've given the airman several chances to change, it's time to write him up.
12. **C (Least effective)** Taking behavior problems to your superior shows you lack command and leadership of your unit. You must learn to handle these problems yourself.
13. **A (Most effective)** Making the airman retake the course ensures he understands the material to do his job safely. It also gives him a chance to mature without adversely affecting his career.
14. **E (Least effective)** Doing nothing risks your unit's safety since the training helps them do their job safely.
15. **B (Most effective)** If you don't have solid evidence of the officer taking advantage of others, you must submit to his directive. Learning how to settle disagreements without involving superiors is a leadership trait you must cultivate.

16. **C (Least effective)** Disobeying a commanding officer's orders is insubordinate. Expect disciplinary action.
17. **C (Most effective)** If you have a problem with someone, the responsible action is to schedule a meeting and talk it through. You're assuming the schedule is in retaliation, but it just may be a scheduling mix-up. You don't know until you talk to him.
18. **D (Least effective)** Going to a peer's superior with your suspicions makes you a weak leader. A good leader finds a way to remedy the situation so it satisfies all parties.
19. **B (Most effective)** You must follow orders, but you also need to be impartial. Only document what you physically see.
20. **E (Least effective)** This is not an opportunity to bash your commanding officer. If you have problems with him, the mature way is to talk to him first.
21. **B (Most effective)** Since the new plan affects everyone, talking with your fellow officers helps you all find ways to succeed in the current situation without making excuses.
22. **C (Least effective)** Berating your team for results beyond their control undermines your authority and erodes the respect your subordinates have for you.
23. **A (Most effective)** Scheduling a meeting with your peer officers to discuss the situation allows everyone the opportunity to either rectify the situation or explain what's happening without being accused of fraud.
24. **D (Least effective)** Ignoring a problem doesn't make it go away. The situation needs to be resolved.
25. **C (Most effective)** Since it was your mistake, immediately go back to the base to secure the report before anyone sees it.
26. **D (Least effective)** A subordinate clearly doesn't have the necessary security clearance to see the report. Allowing a subordinate to retrieve it would jeopardize its classification and possibly its mission.
27. **B (Most effective)** Speaking to your superior about workload improvements for your fellow officer is a sign of your leadership responsibility. Once your superior is aware, it's up to him to reallocate the workload.
28. **E (Least effective)** Trying to palm off the officer onto your peers is a weak move and undermines your credibility.
29. **A (Most effective)** It is your fellow officer's presentation, so he has the final decision, but you gave your input as a good leader does.
30. **E (Least effective)** Trying to further your career at the expense of a fellow officer is dishonest and passive-aggressive. It will likely backfire.
31. **C (Most effective)** You always want a clear line of communication with your superior, especially when it comes to rumors. Asking for her advice on the situation may help your superior realize it's reasonable to seek a transfer to a new unit.
32. **B (Least effective)** Denying the rumors is lying, which undermines your integrity and limits your authority and effectiveness as a leader.
33. **C (Most effective)** Your superior must make the final decision, but he also needs your honest opinion. Tell your superior what you've personally noticed and leave it to him to make the right decision.
34. **D (Least effective)** Trying to sabotage someone's career is dishonest and against your core values as a leader and an officer.

35. **C (Most effective)** Your commanding officer wants your honest opinion, so it's important to clearly communicate what you know and what you've seen. Let him decide who is the best candidate.
36. **A (Least effective)** It's not about who you like the most; rather, it's about who is the best candidate for the job.
37. **E (Most effective)** It's your responsibility to maintain command over your subordinates. Do what's right, ignoring the veiled threat from the airman.
38. **A (Least effective)** Doing nothing about blatant insubordination undermines your leadership and authority. Make sure you earn respect and keep it.
39. **E (Most effective)** Always be honest and don't form an opinion until you have all the facts.
40. **C (Least effective)** Talking to the press without having all the facts could jeopardize your career and erode your superiors' confidence in your decision-making skills.
41. **D (Most effective)** Since this project was assigned by your superior, it affects him as well, if it's not done on time. He should have the final decision on whether it's extended or assigned to another officer.
42. **C (Least effective)** Your superior assigned the project to officers, so commanding two subordinates to work on it undermines his orders. Reassignment should be your superior's decision, not yours.
43. **A (Most effective)** Finding innovative ways to overcome challenges is part of leadership. You can retype the report offline, print out a hard copy, and still have it to your superior on schedule.
44. **D (Least effective)** You're lying by omission, which will undermine your superior's confidence in your leadership and decision-making skills.
45. **B (Most effective)** Clear communication with your new commanding officer is of utmost importance. By explaining the situation and asking for help to tackle your new responsibilities, you are demonstrating strong leadership.
46. **E (Least effective)** Sticking your head in the sand and doing nothing will not make the situation go away and will jeopardize your relationship with your new commanding officer.
47. **A (Most effective)** Contacting the responsible party will help you determine how long the wait will be. Then you can explain the situation to the civilians and ask for their patience.
48. **C (Least effective)** Shifting the problem to someone else is not a trait of an effective leader.
49. **B (Most effective)** Your commanding officer deserves the opportunity to address the rumors and remedy the situation in whatever way he feels is appropriate.
50. **A (Least effective)** Berating the airman under your command in such a public way, especially without proof, erodes your authority as a leader by undermining the respect your unit has for you.

SELF-DESCRIPTION INVENTORY

The Self-Description Inventory section of the AFOQT is a personality test. Your answers do not affect your score. This section is not graded. The Air Force wants to determine your personality traits. Since there is no wrong answer, answer the questions without overanalyzing them; your initial reaction to the statement will most likely be the most accurate response. Give your best answer and don't stress out during this section. You will simply be given a series of statements and then asked to determine how well those statements describe your personality.

PHYSICAL SCIENCE

This section assesses your retention of topics learned in high school science classes. While chemistry and physics involve complex equations and reactions, physical science questions are primarily based on your understanding of vocabulary.

Preparation

Understanding scientific terminology is key to learning about physical science. Reviewing the glossary of a physical science textbook is the best preparation, but you can start your review with the terms listed below.

Absolute zero: At 0 on the Kelvin scale (–273.15°C or –459.67°F), this is the lowest temperature attainable and occurs when particles lose the maximum amount of energy.

Acid: A substance with a pH below 7. Acids donate protons and produce hydrogen ions in solutions.

Amplitude: The distance between the crest or trough of a wave from its resting position, or half the distance between the crest and trough of a wave.

Atmospheric pressure: Pressure exerted by Earth's atmosphere. At sea level, standard atmospheric pressure is 100 kPa (kilopascals), 1atm (atmospheres), 760 mm Hg (millimeters of mercury), or 760 Torr.

Atom: An atom is the smallest particle that exists of any given element. It consists of a nucleus (containing protons and neutrons) and the surrounding electrons.

Atomic mass: Measured in atomic mass units (amu), atomic mass is the mass of one atom of an element. The mass of an amu is determined by the mass of 1/12 of a carbon-12 atom.

Atomic number: The identifying number of an element on the periodic table, which is equal to the number of protons in the nucleus. Elements are arranged on the periodic table by increasing atomic number.

Atomic weight: Different than the atomic mass, the atomic weight is the weighted average mass of all isotopes of an element. While the atomic mass of most oxygen atoms is 16 amu, the atomic weight of oxygen is 15.999 g/mol (grams per mole), because a small percentage of oxygen isotopes have a mass less than 16 amu.

Base: A substance with a pH above 7. Bases accept protons and produce hydroxide ions in solutions.

Boiling point: The temperature at which the molecules of a substance have enough energy to switch from the liquid phase to the gas phase.

Celsius scale (°C): One of three main temperature scales. This scale determines the freezing point of water to be zero degrees and the boiling point at 1 atm, to be 100 degrees.

Change of state: Matter can exist in several states or phases including Bose-Einstein condensate, solid, liquid, gas, and plasma. Changing from one of these phases into another is considered a change of state. Example: Freezing a substance from a liquid phase to a solid phase is a change of state.

Compound: Two or more elements chemically bonded together. Example: NaCl (table salt) is the compound created when sodium and chlorine bond together.

Condensation: An exothermic phase change that results in a gas cooling into a liquid.

Conduction (electrical): The process of electrons flowing through conductive matter.

Conduction (heat): The process of heat energy moving from a warmer object to a cooler object through physical contact.

Convection: The transfer of heat through a fluid (liquid or gas) caused by hotter (and therefore less dense) material rising and colder (more dense) material sinking.

Covalent bond: A chemical bond formed when atoms share electrons.

Decay (radioactive decay): Radiation causing the loss of mass. The result of this process is the changing of one element into another. Example: Bismuth-210 decomposes to polonium-210 through beta-emission.

Density: A measure of the closeness of particles in a substance, which is found by dividing the mass of the substance by the volume it occupies. Units of density are typically grams per cubic centimeter (g/cm^3).

Diffusion: The process of particles of two or more substances moving until equally distributed and mixed together. The rate of diffusion is determined by the concentrations of the substances and the energy in the particles.

Dissolve: When components of a substance separate and move throughout a liquid, but do not chemically react with the liquid, they are said to dissolve.

Electric potential: Commonly called voltage because it is measured in volts. Electric potential is the amount of electric potential energy per charge.

Electron: The negatively charged subatomic particle that orbits the nucleus of an atom.

Element: One of about 100 basic substances that cannot be chemically broken down into a simpler substance. The periodic table of elements lists all known elements in order by the number of protons in the nucleus.

Energy: The ability to do work or produce heat.

Explosive: A property that allows for a very quick, almost instant, decomposition of a substance with the release of a large amount of energy.

Force: A push or pull on an object.

Fluid: A substance that has molecules that are able to flow past one another. Fluids include liquids, gases, and plasmas.

Freezing point: An exothermic phase change from liquid to solid. The melting point is the endothermic phase change that happens at the same temperature.

Frequency: A measure of the number of waves that pass a specific point per second. The unit for frequency is the Hertz (Hz).

Gamma rays: Damaging waves of radiation produced by the release of photons from the nucleus of a radioactive element.

Gas/gaseous phase: The phase of matter where molecules have enough energy to break the bonds attracting them to one another. Gases do not have a specific shape and will fill the volume of the container they are in. Gases can be compressed into smaller volumes.

Gravity: The attractive force that exists between any two physical objects that have mass.

Group: Also known as a *family*. Columns on the periodic table are known as groups and share similar chemical properties based on the number of valence electrons they have in common.

Half-life: The amount of time that it takes for a radioactive element to lose half of its radiation.

Heat: The transfer of energy as the result of a temperature difference between a substance and its environment. Heat always flows from higher energy to lower energy.

Heat capacity: The amount of heat needed by a substance to result in a change in temperature. Heat capacity is mass-dependent while specific heat capacity is not.

Heat of combustion: A property of a substance that indicates the amount of heat that is released when one mole of the substance combusts. Each substance has a specific heat of combustion.

Hertz (Hz): The unit of measurement for wave frequency, which measures wave cycles per second.

Ion: A charged atom. Negatively charged ions (anions) are atoms that have more electrons than protons. Positively charged ions (cations) are atoms that have fewer electrons than protons. The movement of electrons during ion formation can cause an electric field.

Ionic bond: A chemical bond formed by the attraction of oppositely charged ions.

Isotope: Atoms of the same element that have a different number of neutrons. For instance, carbon-12 has 6 protons and 6 neutrons while its isotope, carbon-14, has 6 protons and 8 neutrons.

Kinetic energy: The energy of motion.

Latent heat: The amount of heat that a substance will absorb or release during a phase change while the temperature remains constant.

Lens: A transparent object with a curved surface that causes light to bend.

Liquid/Liquid phase: The phase of matter where molecules have enough energy to move past one another, but not break the bonds attracting them to one another. Liquids do not have a fixed shape but do have a fixed volume.

Mass: A measure of the amount of matter in something. Mass is different than weight (which is a measure of mass multiplied by the effects of gravity), but the terms are often used interchangeably by non-scientists.

Matter: Any substance or object that has both volume and mass.

Medium: The material through which a wave travels.

Melting point: An endothermic phase change from solid to liquid. The freezing point is the exothermic phase change that happens at the same temperature.

Metal: An element that releases electrons and creates a positive ion (cation). Metals are good conductors of electricity and heat. They are known for being malleable and ductile, and have a metallic luster. The cations of transition metals exist in a sea of free-floating electrons, which gives them their unique properties.

Mixture: A combination of two or more substances that can be separated by physical means.

Mole: 6.022×10^{23} of any item. Massive amounts of particles are measured in moles. The number 6.022×10^{23} is known as Avogadro's number, named after the scientist that theorized its existence.

Molecule: The smallest group of chemically-bonded atoms that can exist and still react and retain the properties of a substance.

Neutron: A neutral (no charge) subatomic particle that exists inside the nucleus of an atom.

Newton (N): A unit of force named after Sir Isaac Newton. One newton = 1 kilogram meter per second squared (1 kg m/s^2).

Noble gases: A group of elements in the far right column of the periodic table, known for being unreactive.

Nucleus: The center of an atom which contains the protons and neutrons.

Ohm: The units that describe the electrical resistance of a conductor.

Particle: A small piece of matter.

Period: One of eight horizontal rows on the periodic table.

Periodic table: An organizational chart of all of the known elements in order of increasing atomic number (number of protons).

pH: A measurement of the hydrogen ion concentration in a solution.

Phase: A state of matter dependent on the amount of energy in a substance's molecules. The most common phases are solid, liquid, and gas.

Photon: A particle with zero mass that transmits light or other electromagnetic radiation.

Potential energy: The energy an object holds due to its position relative to other objects.

Pressure: The amount of force exerted per given area, measured in pascals (Pa).

Proton: A positively charged subatomic particle in the nucleus of an atom. The number of protons in a nucleus determines an element's atomic number.

Radiation: Heat energy moving from a warmer object to a cooler object through waves. It is the only form of heat transfer which does not require a medium.

Solid/solid phase: The phase of matter where molecules do not have enough energy to move past one another, which forces a fixed shape and volume.

Wave: An energy-carrying disturbance.

Matter

Everything in the universe can be classified as either energy or matter. *Matter* is anything that has both mass and volume, no matter how small. Energy is everything else. For example, particles in the air are matter because they have mass and take up space. Heat does not have mass or take up any space, so it is energy.

Classification of Particles of Matter

The smallest particle of matter that still retains its distinctive properties is called an *atom*. There are slightly more than 100 different types of atoms. These are called elements and are organized by the number of protons in their nuclei on the periodic table of elements. Atoms are the basic building block of all matter.

Chemically bonded groups of atoms are called *molecules*. Molecules can be different types of atoms that are bonded together, such as carbon dioxide (CO_2), or they can be groups of atoms of the same element bonded together, such as the majority of oxygen in our air (O_2).

When an atom loses or gains negatively charged electrons, it becomes positively or negatively charged itself. A charged atom is called an *ion*.

Atomic Structure

Models

In order to study an item that is too small to be visibly seen, scientists must develop models based on observed effects of the presence of that item on its environment. As more experiments are conducted and technology advances to allow for even more observations, the model improves over time. The model of the atom is one such model; it has developed over time as scientists have learned more about the atom's effects.

While Democritus developed the idea of the atom thousands of years ago, the usable model of the atom was not developed until about two hundred years ago by John Dalton. It wasn't developed further for nearly another hundred years, when scientists J.J. Thompson, Ernest Rutherford, and Neils Bohr were all able to contribute to the complexity of the model. The most common model studied has been the Bohr model, which resembles a solar system with a center (nucleus) and orbiting bodies (electrons).

While the Bohr model is very useful for learning basic atomic structure, more current models are more accurate in showing the existence of electron clouds instead of perfect planes in which the electrons orbit.

Subatomic Particles

There are three subatomic particles that make up most atoms. *Protons* are positively charged and located in the nucleus at the center of the atom. They have a mass of 1 atomic mass unit (amu). *Neutrons* are neutrally charged particles that exist with the protons in the nucleus of an atom. Neutrons also have a mass of 1 amu. Lastly, the *electrons* are negatively charged particles that orbit the nucleus. They are much smaller in terms of mass, nearly 2000 times less massive than a proton or neutron.

Atomic structure

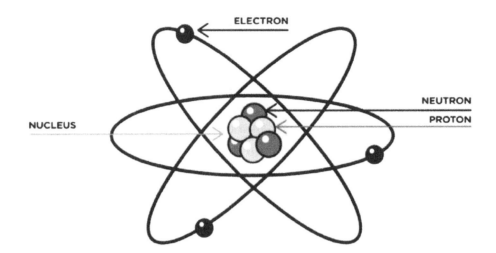

Atomic Number, Atomic Mass, and Isotopes
The number of protons in an atom determines many of the properties of an atom and identifies which element an atom belongs to. For this reason, the *atomic number* of an atom is equal to the number of protons in its nucleus. Elements on the periodic table are arranged by increasing atomic number.

The *mass number* of an atom is equal to the number of protons added to the number of neutrons. The mass of an electron is so tiny that it is not considered when determining the mass number of an atom.

All atoms of the same element have the same number of protons. The neutrons are usually about the same but can vary some. Atoms of the same element with different numbers of neutrons are called *isotopes*. Because the mass of neutrons is considered in an atomic mass, isotopes of the same element will have different atomic masses. The atomic weight shown on the periodic table of elements is the weighted average mass of all isotopes of an element. Different isotopes of the same element will still have the same atomic number because the number of protons does not change.

Electron Configuration

The arrangement of electrons orbiting a nucleus is very specific and called an *electron configuration*. Electrons are negatively charged, while protons are positively charged. Opposite charges attract, so electrons like to be as close to the protons in the nucleus as possible. However, there are specific places an electron is allowed to be located in its orbit of a nucleus. There are eight shells around a nucleus that can be occupied by an electron. Each of these shells has between one and four subshells. Each subshell has between one and seven orbitals, and each orbital can hold two electrons. The subshells of an atom each vary in their distance from the nucleus. Electrons will fill in the orbitals of the closest subshells before occupying an outer subshell.

The way electrons are positioned around an atom determines its chemical reactivity and is important to understanding how atoms of specific elements will bond with other atoms of another element. Because of this importance, the elements on the periodic table are arranged in horizontal rows called *periods* based on the shell that the outermost electron occupies. They are also arranged into vertical columns called *groups or families* depending on the number of electrons in the outer shell that is available to form a chemical bond. These outer electrons that are able to form bonds are called *valence electrons.*

When an outer electron shell is completely filled up and the next shell is still empty, an atom will not bond with others. These atoms are called noble gases and occupy the far-right column of the periodic table. Noble gases include helium (He), neon (Ne), argon (Ar), krypton (Kr), xenon (Xe), and radon (Rn).

Classification of Matter into Elements, Compounds, or Mixtures

An *element* is one of about 100 basic substances that cannot be chemically broken down into a simpler substance and still have the same properties. The periodic table of elements lists all known elements in order of the number of protons in the nucleus. Most of these elements are naturally occurring and have been on our planet from the time it was created. However, elements above Uranium (92 on the periodic table), have been created in a lab.

A *compound* is made up of two or more elements chemically bonded together. Some compounds are simple and contain only two elements, like table salt, which is made with a single bond between a sodium ion and chloride (NaCl). Other compounds are incredibly complex, containing several elements and hundreds of thousands of chemical bonds, like a protein found in human muscles.

Chemical compounds always have constant proportions of each element by mass. For instance, formaldehyde (CH_2O), a compound that is dangerous for humans, is composed of 1 atom of carbon, 2 atoms of hydrogen, and 1 atom of oxygen, a 1:2:1 ratio. Glucose ($C_6H_{12}O_6$), our body's primary source of energy, is composed of 6 atoms of carbon, 12 atoms of hydrogen, and 6 atoms of oxygen. The compound glucose is made of the same elements that simplify to the same 1:2:1 ratio as the compound formaldehyde. However, instead of poisoning our bodies, glucose is needed by our bodies. These two compounds have the same elements in the same ratio but have different proportions of each element by mass. This difference is all it takes to make them very different compounds.

A mixture is made up of two or more elements, but unlike a compound, the elements can be physically separated. Two common physical separation methods include filtration and evaporation. Gravel and water in a bucket would make a mixture that could be filtered in order to separate it into its components. Salt dissolved in water would make a mixture that could be separated by using evaporation.

Mixtures can be further broken down to be classified as either homogeneous or heterogeneous mixtures. *Homogeneous* mixtures are ones that are uniform throughout the mixture. Any given sample of the mixture will have the same ratio of the mixture's components. A homogenous mixture where one of the components is a liquid is called a *solution*. Salt dissolved in water is a common solution. All solutions are homogeneous mixtures, but not all homogeneous mixtures are solutions.

Heterogeneous mixtures are not evenly mixed throughout. Samples taken from the same heterogeneous mixtures could show a variety of ratios of constituent components when compared to one another. The example of gravel in water is an example of a heterogeneous mixture.

Categorization of the States of Matter

There are five physical states of matter: *Bose-Einstein condensate, solid, liquid, gas, and plasma.* While plasma is the most common state of matter in the universe, it is not as common on Earth. *Bose-Einstein condensate* is a very uncommon state of matter; it is only created in a lab when the temperature of something nears absolute zero. So the three most common states of matter on Earth are solid, liquid, and gas.

States of matter are determined by the energy in the molecules that make up a substance. At lower energy levels, molecules are more likely to be strictly bonded into rigid positions, where at higher energy levels, molecules are not bonded to one another and are free to travel away from one another. The three states of matter (solid, liquid, gas) are categorized by the effects of this energy in two ways: the effects on the substance's volume, and the effects on the substance's shape.

Solids have lower-energy molecules that bind them into a fixed shape and volume. Solids cannot be compressed.

Liquids have more energy in their molecules, which gives them the flexibility to move past one another. The result is that liquids do not have a fixed shape. They simply take the shape of the container they occupy. The molecules do not have enough energy to change their volume; liquids have a fixed volume and are not able to be compressed.

Gas molecules have enough energy that they are not held together by their bonds. Gas molecules can move freely and will spread out to fill the space they occupy. For this reason, gases do not have a fixed shape or volume, and they can be compressed. Specific gas laws govern how the volume of gases are affected by outside forces such as temperature and pressure.

Relationships with Matter and Energy

Thermodynamics is the study of heat and its relationship with other forms of energy, work, and temperature. There are four laws of thermodynamics, but the first two are the most common.

Conservation of Matter
All matter is conserved. The *law of conservation of matter* states that matter cannot be created nor destroyed. It can only be rearranged. The basic elements will always stay the same, but their bonds to one another can be broken, and they can be reorganized into other items. For instance, when you burn a log, you are left with ash. It may seem as if the matter was destroyed, but the mass of the log before it was burned, and all of the ash as well as the atoms that were released into the air as a gas, collectively have the same mass.

While the conservation of energy and conservation of matter are often studied as two separate laws, Einstein's famous formula $e=mc^2$ shows a way for mass and energy to form from one another. The two separate laws are really part of the same law: *the law of conservation of mass-energy.* It is helpful to think of the law of conservation of mass-energy as two separate laws unless you are actually trying to convert mass into energy.

First Law of Thermodynamics - Conservation of Energy
Known as the *conservation of energy*, the first law of thermodynamics states that energy cannot be created or destroyed, only transferred. When you hold an ice cube in your hand, your hand becomes cold and eventually the ice cube melts. A common misconception is that the ice cube transfers its cold into the hand, but cold isn't a type of energy or anything that can be transferred. Cold is merely an absence of heat. What really causes the feeling of cold is the absence of heat as the energy from the hand moves into the ice cube. Heat is never lost, only transferred.

Sometimes other types of energy are transferred into heat. For example, using energy from muscles in your arms to rub your hands back and forth together briskly creates friction, which causes heat. The energy put into rubbing hands together is ultimately converted into heat energy through friction.

All friction results in heat energy, but it doesn't *produce* heat energy. It is simply converting the energy introduced to the system into heat energy. Because even the air causes friction, commonly called drag, this law explains why perpetual motion cannot truly exist. No matter how smooth a machine is built, there is always a source of friction which will convert a bit of the energy of motion into heat until all of the energy has transferred away from the system.

Different Forms of Energy
Common forms of energy include: chemical, thermal, radiant, nuclear, elastic, gravitational, and mechanical. Energy can transfer from any of these forms to another within a system, but also between systems. One system transferred energy to another. An apple on a tree uses radiant energy from the sun to grow. That energy is stored as chemical energy in the bonds of the apple until you buy it at the store and eat it. That energy is then used by your body for its energy needs.

Kinetic and Potential Energy
Energy can be categorized as the energy of movement, *kinetic energy,* or the energy of position, *potential energy.* Any object that is moving has kinetic energy, while any object that has height or is stretched has potential energy. Potential energy is stored energy that will begin to transfer to kinetic energy as soon as the object starts moving.

Example 1: A book sitting on a table has potential energy that transfers to kinetic energy when the table is bumped and the book falls to the ground.

Example 2: A bowstring pulled back has potential energy that is transferred into kinetic energy in an arrow as an archer lets go of its stretched position.

The formula for kinetic energy is:

KE = ½ mv²   *m* = the mass of the object
                                         v = the velocity of the object

The formula for potential energy is:

PE = mgh   *m* = equals the object's mass
                                         g = equals acceleration caused by the gravitational force acting on the object (9.8 m/s², on Earth)
                                           h = the height of the object above the ground

The total energy for a system is found by adding the potential energy and the kinetic energy together. Because energy cannot be gained or lost, kinetic and potential energy transfer back and forth.

$$\text{Total Energy} = KE + PE$$

A skateboard on a hill uses this transfer of energy back and forth between potential energy and kinetic energy. Work is put into the skateboard as the rider rolls the skateboard to the top of the first hill. At this point, all of the energy that was put into the skateboard is converted into potential energy. As the skateboard travels down the first hill, that energy is transferred into kinetic energy and it increases in speed. At the bottom of the hill, the skateboard is going very fast, as all of the energy becomes kinetic energy. This energy is used to get the skateboard up the next hill, as the kinetic energy is transferred into potential energy during the incline. Because some energy is lost to friction with the ground, the next hill must be smaller as the skateboard continues on its path until it needs to be pushed again by its rider.

Kinetic Energy Skateboard

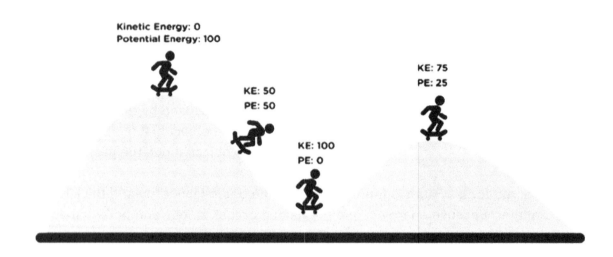

Transferring Thermal Energy
Conduction, convection, and radiation are all ways that thermal energy (heat) is transferred. The main difference between the three is the medium in which the heat transfers. Conduction transfers heat through a solid. When two objects are physically touching one another, the heat energy can flow from the molecules of one object to the molecules of another object. This is what happens when an ice cube melts in a hand and is why a metal spoon in a hot pan will have a hot handle after sitting there for a couple of minutes.

Convection transfers heat through a fluid. Fluids include both liquids and gases. Heat is still transferred from molecule to molecule. The closeness of the molecules varies depending on the state of the fluid. The molecules in liquids are all touching, so the heat can transfer rather quickly. The molecules in gases come into contact with one another and transfer heat when they do, but they are not in constant contact, so the heat will travel a bit more slowly. Increasing the speed at which the molecules are moving will encourage more interactions. Using a fan is one way to do this.

Radiation is the only method of transferring thermal energy that doesn't require matter as a medium. While conduction and convection transfer heat through molecules, radiation uses waves to transfer heat. We feel the sun's heat as *electromagnetic waves* that have traveled through the vacuum of space before reaching us.

Measuring the Transfer of Thermal Energy

When transferring heat through convection or conduction, the thermal energy is transferred between molecules. Because the molecular bonds in every material are different, every material transfers heat differently. Some materials are more resistant to transferring heat than others. That's why using a wooden spoon to cook will cause fewer burns than a metal spoon will. Wood does not transfer heat as well as metal. The *specific heat capacity* of a material is the amount of heat that is needed to increase the temperature of one gram of material by one degree Celsius or Kelvin, assuming pressure is kept constant. When the specific heat capacity of a material is known, the following equation can be used to determine how much heat is needed to warm it up to a specific temperature:

$q = s \times m \times \Delta t$

q = the amount of heat transferred
s = the specific heat capacity of the material being used
m = the mass of the material being used
Δt = the change in temperature

Calorimetry is the technique of measuring the amount of thermal energy transferred. Scientists use an apparatus called a calorimeter to help them with this in labs, but a simple calorimeter can be made with a Styrofoam coffee cup, lid, and thermometer inserted through the top of the lid.

Temperature Scales

When measuring temperature, scientists generally use the Celsius scale in their experiments, but there are actually three different temperature scales that can be used. In the United States, the most common temperature scale is the Fahrenheit scale. The Fahrenheit scale uses 32°F as the point at which water freezes, and 212°F as the point at which water boils. In the rest of the world, the most common temperature scale is the Celsius scale. The Celsius scale uses 0°C as the point at which water freezes, and 100°C as the point at which water boils. The last temperature scale is actually the accepted SI (Système international d'unités) measurement for temperature and is called the Kelvin scale. The distance between degrees is the same as the Celsius scale, but the Kelvin scale uses absolute zero (-273.15°C) as 0K (kelvins).

The following formulas can be used to convert temperature from one scale to another:

Fahrenheit to Celsius: °C = 5/9(°F - 32)

Celsius to Fahrenheit: °F = 9/5(°C) + 32

Celsius to Kelvin: K = °C + 273.15

Second Law of Thermodynamics - Law of Entropy

Everything in the universe is moving toward entropy (disorder). The *law of entropy* states that entropy, or disorder, will always increase in a natural system. Heat energy will always disperse and move from warmer to cooler until it is spread out. A hot cup of cocoa will always cool if left sitting on the counter. The heat will disperse onto the counter and into the air. If the cup of cocoa were to stay hot indefinitely, this would violate the second law of thermodynamics.

Work is done on an object to transfer energy to it. In the case of making cocoa, a microwave may convert electrical energy coming into the house from a power company to heat the cocoa. The power company may use the chemical energy stored in coal to create the electrical energy.

Chemical Changes vs. Physical Changes

Matter undergoes either chemical changes or physical changes in order to transform. Physical changes don't involve the chemical composition of matter changing. For instance, crumpling a sheet of aluminum foil into a ball does not make it any less aluminum. Another example would be a phase change. Water in the form of ice is not chemically different than water in the form of steam. The molecules still have the structure of water. They are still chemically water; they are just in a different phase.

A *chemical change,* on the other hand, changes the chemical composition of matter. Iron rusting, bread molding, or a log burning are all examples of a chemical change. The atoms and molecules in each case have separated or bonded with other elements in the environment to become a chemically different object. Chemical changes are much more permanent than physical changes which can, many times, be reversed.

Natural Occurrences of Energy and Matter Conservation

There are examples of the conservation of energy as well as the conservation of mass all around the natural Earth.

The sun is the primary source of energy for the living organisms on Earth, but while plants can use the sun to produce energy for life, animals cannot. Animals, including humans, obtain the energy needed to live by consuming plants, which transfer the sun's energy into chemical energy. Eating a plant makes that energy available for the animal to use to power its daily needs. Animals can also eat other animals and use the energy stored in the other animals' bodies. This transfer of energy through the food chain is divided into *trophic levels*. Plants occupy the first trophic level as producers. Herbivores occupy the second trophic level and are able to receive a plant's energy when they eat it. Omnivores and carnivores occupy the higher trophic levels. As the transfer of energy continues up each trophic level, only 10% is used by the next organism. Energy is transferred even in the decomposition of plants and animals.

Energy and matter transfer can also be seen in the transformation of geological formations. Igneous, metamorphic, and sedimentary rocks are transformed into one another through heat transfer happening over time. Heating, cooling, pressure, weather, and erosion all play a part in these changes and are all in response to energy transfers. Below Earth's crust, radioactive elements produce energy as they decay. This heat warms magma, which then causes Earth's plates to move, changing currents in the ocean. This movement of energy causes all of the natural events that play a part in the transformation of these rocks over time.

Characteristics of Radioactive Materials

Isotopes and Radioactive Decay
Isotopes are atoms of the same element that have a different number of neutrons in the nucleus. Some isotopes have nuclei that are quite stable, but others are not. When the nucleus of an atom is unstable, it will break down by releasing particles or energy as radiation. This is called *radioactive decay*. Isotopes of larger elements are more likely to experience radioactive decay than isotopes of smaller elements.

As isotopes decay, they decrease in size and lose enough particles that the number of protons changes and they change into a different element. This happens at a regular rate for each isotope. The amount of time for the mass and radiation of a specific isotope to deteriorate into half of its original amount is called its *half-life.*

As isotopes decay, they become isotopes of other elements, known as *daughters*. If the daughter of a decayed isotope is also unstable, it will decay even further. As soon as an isotope decays down to a stable daughter, the radioactive decay will stop.

Alpha Particles, Beta Particles, and Gamma Radiation
As mentioned before, radioactive decay involves releasing tiny particles or energy. *Gamma* radiation is the most dangerous and involves the release of photons of energy. *Alpha* radiation releases a particle called an *alpha particle* that has two protons and two neutrons bonded together. *Beta* radiation releases *beta particles*, which are high-speed electrons or positrons.

Nuclear Reactions
There are two types of nuclear reactions, and they both produce energy. *Fission* is the breaking apart of an atom and creates the energy made in nuclear power plants. This energy is created by bombarding unstable isotopes with high-speed particles.

While plenty of energy can be created in nuclear power plants with the fission of atoms, it is nothing compared to the energy created by the sun during the fusion of atoms. *Fusion* is the combining of the nuclei of two smaller atoms to create a larger atom.

Physics

Mechanics

Mechanics, one of the main branches of study in physics, is the study of the effects of energy and forces on objects. Within the field of mechanics, *dynamics* is the study of forces on moving objects, *statics* is the study of forces on objects that are not moving, and *kinematics* is the study of an object's motion without regard to outside forces.

<u>Forces</u>
It is common to use the term *weight* when what is really measured is mass. Weight is a force, measured in newtons. Mass is a measure of how much matter an object has. The mass of an object is the same no matter where it is located, but weight is dependent on the amount of mass times the acceleration due to gravity, so the weight of an object is different on different planets. The acceleration due to gravity on the moon is one-sixth of the acceleration due to gravity on Earth, so an object on the moon will weigh one-sixth the amount that it does on Earth. Newton's second law gives us the formula for weight:

$$\text{force} = \text{mass} \times \text{acceleration}$$
$$\text{or}$$
$$\text{weight} = \text{mass} \times \text{acceleration}$$

Friction is a force that works on an object in the direction opposite of its motion. Its strength is dependent on surface irregularities between the object and anything it touches. Any matter can cause friction on an item, even air. Even before an item moves, it must first overcome the *static friction* that exists when it is in contact with another surface. We use friction to our advantage when we walk. Think of different surfaces that you have walked on. Surfaces with more irregularities are easier to walk on because there is more friction. For example, a carpeted floor is easier to walk on than an icy pond.

Two terms that are frequently misunderstood are *centripetal force and centrifugal force.* When an object is spinning around an axis, the force that keeps it spinning and not flying off on a linear path tangent to the orbit of the spin is called the centripetal force. A centripetal force is not a force in and of itself, but a label for any force that keeps an object in its rotational path and is directed toward the center or axis. A centrifugal force is a *fictitious* force. It isn't part of a force interaction. What is really meant by a centrifugal force is the lack of a centripetal force. For instance, when your clothes are spinning in the washing machine, there is a centripetal force acting on the clothing and keeping it in rotation. The centripetal force is not acting on the water that flies through the holes in the drum of the washer. There is not a centrifugal force acting on the water, just the lack of a centripetal force.

Besides the three laws of motion in the next section, Isaac Newton also gave us the *law of universal gravitation* based on his discovery that gravity is universal. The law of universal gravitation tells us that there is a gravitational force between every object in the universe. This force is directly proportional to the masses of the objects being considered and inversely proportional to the square of the distance between those objects. This means that if two objects are the same distance away from you, but one is twice as massive than the other, the gravitational pull between you and that object is twice as strong. However, if two objects have the same mass, but one is twice as far away from you as the other, the gravitational pull is only one-fourth of what it is between you and the closer object. The law of universal gravitation can be shown by the formula below:

$F_g = G m_1 m_2 / r^2$

F_g = gravitational force
G = the universal gravitation constant, 6.67×10^{-11} N m²/kg²
m_1 = mass of the first object
m_2 = mass of the second object
r = distance between the two objects

Newton's Laws of Motion

The most important work with regards to calculations and motion are the three laws of motion observed and described by Sir Isaac Newton. These are known as Newton's laws of motion, and all involve the vector quantity *force*. A force is defined as a push or pull. So important was the work of Isaac Newton that the internationally recognized unit of measurement of force is the newton (N).

Newton's First Law of Motion

The first law of motion is sometimes referred to as the law of *inertia*. Inertia is a concept that was developed before Newton's time. This concept was first published by Aristotle and then later refined by Galileo as a measure of an object's need for a force in order for there to be a change in its movement. Isaac Newton published this as the first law of motion more specifically as: *An object in motion stays in motion and an object at rest stays at rest unless acted upon by an outside force.*

An example is an asteroid flying through space in the same direction and at the same speed forever or until it hits another item.

Newton's Second Law of Motion
Newton's second law of motion is best shown by the simple formula below:

$$F = ma$$
or
$$a = F/m$$

The two arrangements of this formula give us three relationships that are described by this law. The first states that a force is equal to the product of the mass of the object exerting the force and the object's acceleration. The units are newtons (N), which are equal to kilogram meters per second squared (kg m/s^2). The second relationship can be seen in the rearrangement of the formula, which shows that a change of force on an object results in a directly proportional change in its acceleration. For example, hitting a baseball with five times the amount of force as it was hit before, results in an acceleration in the baseball that is five times greater than it had before. This rearrangement also shows that mass has an inversely proportional relationship with acceleration. For example, if you hit two balls with the same force, but one ball is three times as massive, the more massive ball will have one-third of the acceleration as the other ball.

Newton's Third Law of Motion
This law states that *for every action there is an opposite and equal reaction* and describes the transfer of force when two items collide. For example, when a bug hits the windshield of your car, your car hits the bug back with the same amount of force, but in the opposite direction. A force acting on a small object causes more damage than the same force acting on a massive object.

Motion in One Dimension
Measurements and calculations in physics are either in *scalar* quantities or *vector quantities.* Scalar quantities involve only a magnitude, while vector quantities include a magnitude and a direction. Common scalar quantities include time, area, and speed. Common vector quantities include displacement, velocity, momentum, and acceleration. When using vectors, the direction can be specifically stated but is more accurately denoted with a positive or negative sign in front of the quantity. This reflects the position of an object in comparison to a reference point. A negative sign in front of the number does not mean a negative quantity; it means that the motion is in the opposite direction of what is referenced. For instance, to compare the motion of two objects, we could say that one is moving 5 mph (miles per hour) north and the other is moving 5 mph south. These descriptions are both vectors (specifically velocity) comparing magnitude and direction of the movement. While the objects both have a speed of 5 mph, they have opposite velocities because they are moving in opposite directions. Another way to show

this information would be to say one object has a velocity of +5 mph and the other has a velocity of -5 mph. The differing signs let us know that the objects are moving in opposite directions, but not the exact directions unless a reference point has been defined.

Velocity, acceleration, and *momentum* are all vectors commonly used when studying kinematics. Of these three, velocity has the simplest description and is the only one that is not reliant on another. Velocity is found by dividing an object's *displacement*, or the change in distance (Δx), by the change in time (Δt). The standard units for velocity are meters per second (m/s). So the formula for finding velocity looks like this:

$$v = \Delta x/\Delta t$$

Acceleration occurs when there is a change in velocity. As a vector quantity, velocity is more than just speed; it is speed with direction. For this reason, there are three ways an object can experience acceleration: speeding up, slowing down, or changing direction. An object going around a curve or turning while maintaining a consistent speed is accelerating, in terms of physics. Because velocity is direction dependent, a change in direction changes velocity. A change in velocity causes an acceleration. To solve for acceleration, initial velocity is subtracted from final velocity and divided by the length of time over which the change occurs. The standard units for velocity are meters per second squared (m/s^2). This is sometimes written as meters per second per second (m/s/s). So the formula that solves for acceleration look like this:

$$a = \Delta v/\Delta t$$
$$\text{Or}$$
$$a = (v_2-v_1)/\Delta t$$

Projectile Motion
Projectile motion is also known as a study of two-dimensional motion because both horizontal and vertical components are considered for projectiles, such as a ball being kicked or a car racing off a cliff. Most of the time, projectile motion is parabolic in shape, but objects in free fall that have no initial horizontal velocity are also considered to have projectile motion. The influence of gravity on an object's vertical motion is independent of horizontal motion, so it acts the same for objects in free fall as it does for objects that are thrown, kicked, or launched.

Momentum
Momentum is another measurement that is dependent on an object's velocity. Momentum can be thought of as a measure of how difficult it is to start an object into motion or to stop an object already in motion. It is also commonly thought of as a measure of an object's inertia. Momentum is denoted by the letter p in equations and is the product of an object's mass and velocity, as shown in the equation below. The standard units for momentum are kilogram meters per second (kg m/s).

$$p=mv$$

p = momentum
m = mass
v = velocity

Momentum can be transferred between objects that come into contact with one another, and it plays an important role in considering damage that can be caused in collisions. This is why faster moving cars and massive trucks cause the most damage in automobile accidents.

The *law of conservation of momentum* dictates that momentum is never gained or lost, only transferred, and applies to both linear and circular motion (where it is known as *angular momentum*). In terms of linear motion, the most common use of the law of conservation of momentum is in the analysis of collisions. Collisions are considered inelastic if the objects colliding remain connected after the collision, or elastic if they collide and then separate. The sum of the momentums of the objects before the accident must equal the sum of the momentums of the objects after the accident to satisfy the law of the conservation of momentum. For elastic collisions, the formula is as follows:

$$p_1+p_2=p_1+p_2$$
or
$$m_1v_1+m_2v_2=m_1v_1+m_2v_2$$

For inelastic collisions, where the objects are joined after the collision, the formula is as follows:

$$m_1v_1+m_2v_2=(m_1+m_2)v$$

Angular momentum is the momentum an object has when it is spinning around an axis. It is a product of the object's mass, velocity, and radius from the axis. The *conservation of angular momentum* requires that if any of these three variables is changed, an inversely proportional change must occur to the combination of the remaining variables. Because most objects don't spontaneously change mass, the two variables that usually change are the radius and the velocity. This can be observed when a spinning ice skater suddenly increases speed just by bringing his arms closer to the center of the spin.

Systems can have angular momentum as well. The axis for a spinning system should be determined by the center of mass of the system. The formula to find the center of mass for a system is shown below:

$$X_{center of mass} = \frac{m1x1+m2x2}{m1+m2}$$

x = distance from the point of origin
m = mass

Simple Machines

Machines are used to multiply forces or change the direction of forces. There are six simple machines that follow the law of conservation of energy in order to transfer or transform energy through the change in forces. Machines help users by changing the ratio of output force by the machine to input force by the user. This ratio is called the *mechanical advantage*. The greater the mechanical advantage of a machine, the more work is transferred from the person to the machine.

$$Mechanical\ Advantage = \frac{output\ force}{input\ force}$$

$$MA = F_{out}/F_{in}$$

Many machines are designed to manipulate the formula in order to decrease the input force needed by the user. Work is the product of force and the distance over which the force is applied, as shown below. Lengthening the distance over which a force is applied requires an inversely proportional change in the amount of force required. For instance, increasing the distance to be four times greater has the result of decreasing the force needed to one-fourth the original amount. When work is done around an axis, it is called *torque*.

$$w = F \times d$$

6 types of simple machines

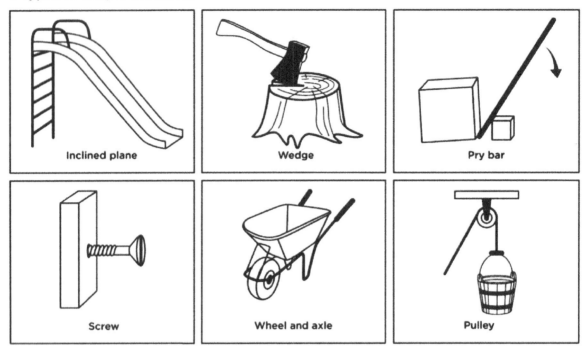

Levers are bars that turn at a fixed point called a *fulcrum*. They increase mechanical advantage by increasing the distance over which work is done, so that less input force is needed. Pry bars are a type of lever.

Wedges work by increasing force for the same amount of pressure applied by the user. Pressure is equal to force divided by area, so wedges decrease the area over which pressure is applied, and the result is an inversely proportional increase in force. An axe is a type of wedge.

Inclined planes work similarly to levers by lengthening the distance over which an object is lifted against gravity. Increasing the distance decreases the amount of input force required by the user. A slide is one example of an inclined plane.

Screws are another type of inclined plane. Working on a much smaller scale, screws are able to greatly increase the distance over which force is applied by wrapping an inclined plane around an axis.

A *wheel and axle* also make use of the rotation around an axis to create mechanical advantage. Wheelbarrows do this to make it easier to haul heavy loads by requiring less input force from the user.

Pulleys also work as a lever around an axis, but ropes are added to change the direction of the force. Multiple pulleys can be used in order to further decrease the magnitude of input force needed by the user.

Electricity and Magnetism

Charge and Electrical Force

Atoms are made up of three subatomic particles: protons and neutrons contained in the nucleus, and electrons orbiting the outside. While neutrons have a neutral charge, protons are positively charged and electrons are negatively charged. Because opposite charges attract and like charges repel, electrons are attracted to the protons in the center of the atom and repel the electrons of other atoms. Generally, atoms have the same number of protons as they do electrons, making them neutral overall, but they can gain electrons and become negatively charged or lose electrons and become positively charged.

Electrical currents are the result of the movement of electrons as they move in reaction to other charges around them. Electrons associated with some materials move more freely than the electrons of other materials. *Conductors* are materials that have electrons capable of moving freely, and *insulators* are materials that have electrons that are much more tightly bound in position and resistant to movement. Electrons in metal, in particular, are loose and not tightly bound to a specific atom. Because of this, metals are very good conductors of electricity.

When two objects rub together, electrons can move from one object to another. The result is the second object carrying a slightly negative charge and the first object carrying a slightly positive charge. This can happen when you slide your feet across carpeting. Electrons are then attracted to you, as you carry a slight positive charge. Receiving a shock as you reach for a light switch is what you feel when electrons jump and move through you to the ground as you reestablish a neutral charge.

Coulomb's Law describes the electrical force that can exist between two particles. It is very similar to Newton's law of universal gravitation, but while Newton's law describes a force between large objects, Coulomb's law describes a force between tiny objects. The following formula is for Coulomb's Law. It shows that the electrical force is directly proportional to the amount of charge in the particles and inversely proportional to the square of the distance between the two:

$F = kq_1q_2/r^2$

F = electrical force
k = proportionality constant, 8.99 x 10^9 N m^2/C^2
q_1 = quantity of charge of the first particle
q_2 = quantity of charge of the second particle
r = distance between the two particles

Electric Current

To have an electric current, a potential difference must be created. A power source, or voltage source, like an electric company, a generator that transfers mechanical energy to electrical energy, or a battery that uses stored chemical energy, creates this difference. This voltage makes it possible for the movement of electrons to create a flow of electricity moving from positive to negative. The amount of current created by a voltage source is measured in amperes, or amps, and is equal to net charge divided by the time it takes to flow, as shown in the formula below:

$I = q/t$

I = electric current in amps
q = net charge in coulombs
t = time in seconds

Electric current can either be *direct current* (DC) or *alternating current* (AC). Batteries use direct current. They have a positive end and a negative end, which allows the current to flow in the same direction at all times. The outlets in your house use alternating current. The flow of electricity changes direction as the polarity of the voltage alternates positions.

The amount of current that flows in a circuit is not only dependent on voltage but also on *resistance*. Resistance depends on the multiple factors that affect the movement of electrons, including the material that the circuit is made of and the temperature. Resistance is measured in ohms and is found using the formula below, which represents *Ohm's law*.

$R = V/I$

R = resistance measured in ohms
V = voltage measured in volts
I = current measured in amperes

Another factor that affects resistance is how components in an electrical circuit are wired. If a single wire connects all of the components along one path, they are wired *in series*. Components wired in series are dependent on the component before to carry the voltage through the circuit. If one component fails, the voltage will stop and the circuit will cease. The following formula shows the calculation for resistance in series:

$$R_{total} = R_1+R_2+R_3+....+R_n$$

If multiple components are connected to the same two points with separate wires, they are wired *in parallel*. Because the components of parallel circuits are not wired together, they are independent of each other and can still continue in the circuit if one fails. The components in parallel circuits split the voltage evenly between them. The following formula shows the calculation for resistance in parallel:

$$R_{total} = (1/R_1)+(1/R_2)+(1/R_3)+....+ (1/R_n)$$

Electromagnets

Ferromagnetic materials, like iron, create a field around them that attracts or repels other ferromagnetic materials. These materials are called *magnets*, and the field is called a *magnetic field*. Magnets have a positive end, or pole, and a negative end, or pole. Sometimes these are referred to as the north and south poles of a magnet. Between any two magnets, a force exists across the magnetic field. This means the magnets do not have to touch to attract or repel each other. When like poles are pointed toward each other, this force will push them apart. When opposite poles are pointed at one another, this force will pull the magnets together.

Electricity can increase the force between magnets. This is how magnets can be made strong enough to lift entire cars. Wrapping a wire around a magnet and sending an electric current through that wire can create a very strong magnet called an *electromagnet*. Electromagnets are useful because they can be custom created to be specific strengths, and their magnetic field can be turned off and turned back on by controlling the flow of electric current through the wire.

Waves

Properties of Waves

Energy is carried on waves from one location to another. The types of waves that carry different types of energy vary, but all waves have several properties in common. Waves all have a measurable wavelength. This is the length of one oscillation of the repetitive motion through which it travels. Waves also have an amplitude. This is the distance from the origin of the wave to its crest (highest point) or trough (lowest point) and is a measure of a wave's intensity. All waves have a frequency, which is dependent on how many wavelengths pass a point during a given time. Lastly, waves have a measurable speed.

Gravitational Waves

A recently discovered wave is called the *gravitational wave*. This wave carries *gravitational radiation*, a type of radiant energy, as it travels along the curvature of space/time. Gravitational waves move as a ripple moving outward from a disturbance. Throwing a pebble in a pond and watching the waves ripple outward from where it hit the water is a way to visualize the motion of a gravitational wave.

Wave-Boundary Interactions

Boundaries, like objects or a change in the medium through which a wave is traveling, can change the movement of the wave in different ways. If a wave moves from one medium to another, the change in matter will change the resistance put on the wave, and its speed will change. This changing of the speed causes the wave to change direction. This direction change is called *refraction* and is based on the density of the medium it is traveling through. Try to catch an object moving in water when you are viewing it from outside of the water and you will experience frustration, as the object is never quite where you think it is. This is due to the refraction of light as it hits the water. It makes the object appear to be in a slightly different spot than where it really exists.

Lenses are either found in nature or created in order to refract light in specific ways. Your eyes have lenses, but *microscopes* and *telescopes* are both tools that were created by using lenses to help us see things that are either too small or too far away to see with the lenses in our eyes.

Prisms are another tool created to utilize refraction. They take advantage of *dispersion*, which is a type of refraction that bends light differently based on the wavelength of light. Prisms cause white light to spread into all the different colors of light as they refract at different angles based on their wavelengths.

When a wave hits a medium that it cannot go through, some or all of it can bounce back. This is called *reflection* and is how mirrors show you your image. *Plane mirrors,* like the one in a bathroom, reflect back all of the light bouncing on them and allow you to see an image of yourself. *Curved mirrors* can be used to manipulate how we see images that are reflected back at us. Curved mirrors can enlarge or miniaturize an image or make objects appear closer or farther away than they really are.

Some objects absorb some wavelengths of light but reflect back others. That is how we experience colors. White light hits an object, and the colors that are reflected back are what we see. Any color that we don't see represents the wavelengths of light that were absorbed. When you look at a green blade of grass, every wavelength of light was absorbed except the wavelength that you perceive as green. That particular wavelength is being reflected back at you.

Total internal reflection happens when a wave is either completely reflected or is unable to leave one medium to enter another. Jewelers take advantage of total internal reflection by cutting diamonds in a way that reflects all of the light that hits them. This makes them appear more brilliant. It is even referred to as the "brilliant cut."

Streaks of light that seem to come through the clouds are an example of another way light can move when it encounters a boundary. *Diffraction* is the bending of light around an object or through a slit.

Light sources emit light waves in random directions. *Polarization* is a technique that creates a selective barrier to alter the motion of approaching waves that are oscillating in a specific direction. Sunglasses use polarized lenses to cut out glare from the reflection of light off of horizontal surfaces.

Wave-Wave Interactions
When waves encounter one another, they interact by either interfering with one another and lessening their effect, or they interact in a way that magnifies their effect. *Constructive interference* occurs when waves are in sync with one another. It is as if they become added together to become one larger wave. *Destructive interference* occurs when waves are out of sync with one another. It is similar to subtracting one wave from another. If there is completely destructive interference, they can even cancel one another out. Interference in sound waves is especially important with musical groups. When members are completely in tune with one another, the sound waves they produce can create a constructive interference that allows them to be louder without the members actually singing or playing more loudly. Destructive interference, however, creates obstacles that make it more difficult to perform and achieve the desired sound.

Mechanical Waves and Sound
Electromagnetic waves do not require a medium and can travel through space even when there is no matter to transfer them, but another type of wave, *mechanical waves,* will only pass through a medium made up of matter. Sound waves are a type of mechanical wave. This is why sound does not travel in space.

Mechanical waves can move in two directions, or a combination of both. *Transverse waves* move at right angles to the direction that the wave is moving. Picking up the end of a spring and moving it up and down or side to side and watching this movement ripple down the length of the spring shows the movement of a transverse wave.

Longitudinal waves move parallel to the direction in which the wave is traveling. Pushing the end of a spring and watching the compression (bunching up) and rarefaction (separating) as it moves down the length of the spring gives a visual representation of the movement of longitudinal waves. Sound waves are a type of longitudinal wave.

When a longitudinal sound wave reaches your ear, your brain translates the vibrations created by the compressions and rarefactions of that wave into a pitch based on the frequency of the wave and intensity (loudness), which is based on the amplitude of the wave. Amplitude is measured using *decibels*. "Deci-" means one-tenth and "bel" stands for the last part of the name of Alexander Graham Bell. A decibel is equivalent to one-tenth of a bel. The decibel scale is a logarithmic scale, so every increase of 10 decibels shows an increase in the intensity of a multiple of 10. For instance, increasing something from 50 dB to 80 dB is a 10^3 increase, so it would be 1000 times more intense.

Musicians use tools like tuning forks, which are made to create a specific frequency of sound wave. The size and material of the tuning fork are carefully chosen so that when it is struck, it will create a sound wave with the exact frequency (in *Hertz,* Hz) of note that the musician wants to hear.

Doppler Effect

The *Doppler effect* describes how an observer perceives light or sound waves coming from a source that is moving. When an object that is producing light or sound approaches a person, the wavelengths appear shorter and shorter. When that object passes and is moving away, the wavelengths appear to increase as the object increases its distance. For sound waves, the pitch of an oncoming siren or another sound will seem to increase as it approaches. As the object making the sound is moving farther and farther away, the pitch will seem to become lower and lower.

The Doppler effect is much harder to detect with objects emitting light, but it does exist. As an object moves closer and closer, its light waves will appear to become bluer as they seem shorter and shorter. As the object moves away, the light waves will seem to increase in length, and the object will give the appearance of becoming more red. This is called *red-shift* and is usually used when describing the motions of stars in the sky. The red-shift of stars in the universe supports the hypothesis that the universe is expanding.

Electromagnetic Waves and Light

When you use your eyes to look around you, you probably perceive that you are looking at objects. You see their shape, details, and different colors. The truth is, our eyes only see one thing: light. When you are in a completely dark room, you can't see the objects in that room. They are still there, but your eyes don't see them. Our eyes only see light and the reflection of light off of the objects it touches.

Light has a dual nature, enabling us to think of it as both moving particles (photons) and as a wave of energy. Waves of light energy travel at an incredible speed of 3.00×10^8 m/s, which is faster than anything we have ever created. The energy in a wave of light is partly electric and partly magnetic, which is why it is considered an electromagnetic wave.

Understanding the nature of light as a wave helps us to understand how our eyes use it to collect information about the world around us.

Light is not the only electromagnetic wave. In fact, light is a very small part of many different types of waves that make up the *electromagnetic spectrum.* The waves with the longest wavelengths on the electromagnetic spectrum are radio waves, whose waves span from just over one millimeter to up to 100 meters in length. Other waves with wavelengths greater than visible light include microwaves, which span between 25 micrometers and 1 millimeter, and infrared waves, which span from 750 nanometers to 25 micrometers. With a span of only 350 nanometers, visible light waves are only a portion of the entire electromagnetic spectrum. The light waves with the longest wavelengths are seen as red, while the light waves with the shortest wavelengths are seen as violet. The colors in between follow the color order of the rainbow (red, orange, yellow, green, blue, indigo, and violet).

Electromagnetic waves that are shorter than visible light include ultraviolet waves that are between one nanometer and 400 nanometers in length, X-rays with wavelengths between one nanometer and one picometer in length, and gamma rays with wavelengths less than a picometer long.

Radio waves, microwaves, infrared waves, ultraviolet rays, X-rays, and gamma rays are all waves on the electromagnetic spectrum along with light waves. Why can we see light and not these other waves? The answer lies in the distinguishing characteristic of the different waves: frequency. Our eyes only perceive electromagnetic waves of a specific frequency. The other waves are all around us; our eyes are just not equipped to see the frequencies at which they travel. In fact, the different colors we see are due to varying frequencies as well. When you are looking at a green apple and a red apple, the reason one appears green to you while the other appears red is purely based on the frequency of light that is reflected toward you as it bounces off the apples. Colors are just our brain's interpretation of different frequencies of reflected light, and it's interesting to think about what colors would exist if our eyes could see electromagnetic waves outside the frequency of visible light.

Chemistry

Periodic Table

<u>Organization of the Periodic Table</u>
The periodic table of elements is a tool that is much more valuable to chemists than just a list of all of the known elements. Every known element is listed in order of atomic number, which is equal to the number of protons in that atom's nucleus.

The unique shape of the periodic table serves an important purpose. The horizontal rows, or periods, represent the different energy levels in the electron cloud. The elements that share a row all have valence electrons in the same energy level. The columns of the periodic table, also known as groups or families, have elements that are similar in chemical reactivity based on the number of valence electrons available for bonding. For instance, the first group on the periodic table is known as the *alkali metals.* Every element in this group is incredibly reactive and is known to explode when exposed to water (except hydrogen). The elements in the second to last group on the periodic table are called *halogens*. These elements are known for creating deadly gases and strong acids. A quick glance at the placement of an element on the periodic table can tell a chemist how large an element is, where the bonding electrons of the atoms of that element are located, and how reactive that element is when brought in contact with other elements.

Periodic Table of Elements

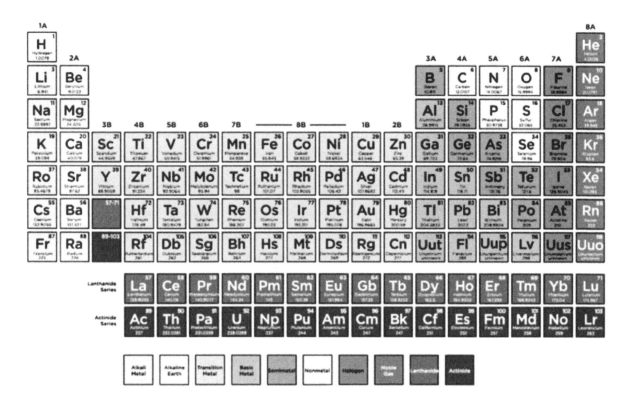

Standard nuclear notation

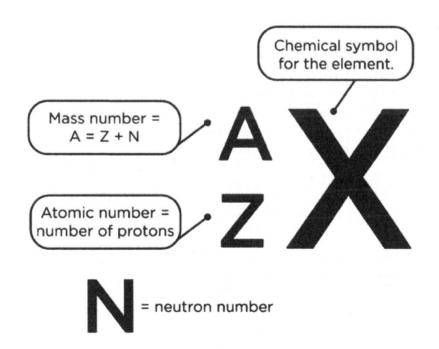

Besides the atomic number, the atomic mass is also listed on the periodic table. The atomic mass is different than the atomic mass number. The atomic mass number is the sum of the number of protons and the number of neutrons in an atom's nucleus. For example, with 16 protons and 16 neutrons, Sulfur-32 has a mass number of 32 amu (atomic mass units). Sulfur-35 has 16 protons and 19 neutrons so its mass number is 35 amu. Atoms of the same element with different numbers of neutrons are called *isotopes.* Most elements have multiple isotopes. The atomic mass listed on the periodic table for sulfur is 32.065 grams/mole. This number is not a whole number because it is a weighted average of all of the isotopes of sulfur that exist in nature.

Along with groups, periods, atomic number, and atomic mass, the periodic table can also be divided into blocks of elements that have similar characteristics. All of the elements fall into one of the three main categories of *metals, nonmetals,* or *metalloids.* Metals are known for being shiny and good conductors of electricity. The elements can be further categorized into smaller blocks known as *transition metals, diatomics, post-transition metals,* and *polyatomic nonmetals.* The two rows of elements listed at the bottom of the main periodic table have special names as well. The row of elements second from the bottom are the *lanthanides,* and the bottom-row elements are the *actinides.* Some of the blocks are named groups like the halogens or alkali metals, as mentioned before. Other groups that create their own blocks are the *alkaline earth metals* and the *noble gases.* The noble gases have a full outer electron shell. This causes them to be mostly unreactive with other elements. They are considered *inert gases*.

The periodic table uses symbols and names to identify each element. The symbol follows one of two guidelines: either it is a single capital letter, or it is a capital letter followed by a lowercase letter. Most of the time, the symbols are derived from the name of the element. For instance, C is the symbol for carbon and Rn is the symbol for radon. Sometimes the symbols are not related to the chemical names as we know them. Some elements have a common name that is listed on our periodic table but was originally named in Latin. For instance, the Latin name for what we know as sodium is natrium. This explains why the symbol for sodium on the periodic table is Na. The name for lead in Latin is plumbum, which is why its symbol is Pb on the periodic table.

Many of the names on the periodic table end in "-ium". While many of the elements ended this way before 1940, all elements discovered after the 1940's were required to end in "-ium". Any element with an atomic number greater than 86, with the exception of lanthanum, ends in "-ium". As new elements are discovered or created, their names are approved by the International Union of Pure and Applied Chemistry, or IUPAC.

Periodic Trends
Along with blocks, rows, and columns of elements that share similar characteristics, the periodic table also shows several trends that allow chemists to easily glance at it and

hypothesize how one element may react to another. *Atomic radii, ionization energy, chemical reactivity,* and *electron affinity* are all characteristics that follow predictable trends. *Ionization energy* is the energy that is required to remove an electron from an atom. Because smaller atoms require more energy to remove their electrons, they have high ionization energy. With the smaller atoms being located at the top of the periodic table, ionization energy decreases from top to bottom. Atoms on the right side of the periodic table have smaller atomic radii than atoms on the left side of the table, so they require more ionization energy as well. For this reason, ionization energy increases from left to right across the periodic table.

As mentioned before, atomic radii decreases from left to right across the periodic table, but they also increase from top to bottom as each row going down the periodic table represents the addition of another electron energy level. Electron affinity represents how easily an atom will gain an electron. With the exception of the noble gases that don't attract electrons at all, electron affinity increases from left to right and from bottom to top of the periodic table.

Chemical reactivity trends diagonally across the periodic table. Metals and non-metals trend in opposite directions with regards to their reactivity. With the left side of the table being metals, the chemical reactivity increases from right to left and from top to bottom. On the right side of the table (with the exception of the noble gases), reactivity increases from left to right and from bottom to top. The top right and bottom left sections of the periodic table contain the most chemically reactive elements. Once again, the exception would be the furthest right column, or noble gases, which are not reactive.

Periodic table trends

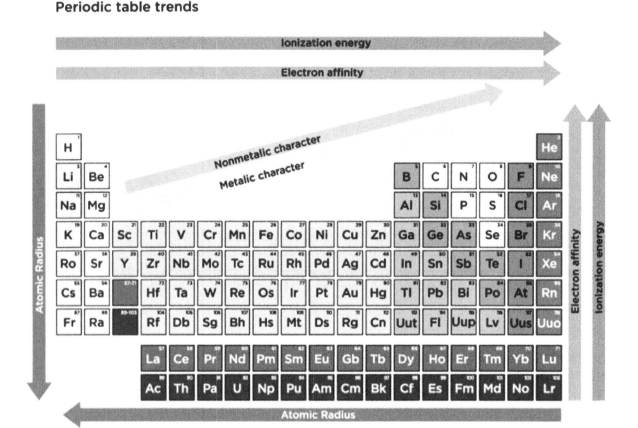

States of Matter

Factors that Affect Phase Changes

The *Kinetic Theory of Matter* explains phase changes and states of matter in terms of energy and motion at the molecular or atomic level. The distance between atoms and molecules determines the state of matter and this distance is dependent on the energy contained in the molecules and atoms themselves. Attractive forces exist between small particles of the same substance. When these particles do not have a lot of energy, they are not able to overcome these forces and will be locked into definite solid shapes and volumes. Particles with more energy are able to loosen the bonds and can slide past one another. They are still locked into a definite volume, but not a definite shape. In a gas, atoms or molecules have enough energy that they can completely overcome the attractive forces between them. They are not locked into a definite shape or volume and will travel to evenly fill any space that contains them.

Temperature is a measure of the amount of average molecular kinetic energy in a substance. This explains why substances are more likely to be gases at a higher temperature and more likely to be solids at a cooler temperature. Higher temperatures mean the particles have more energy to overcome the attractive bonds that would hold them together.

Pressure and the Ideal Gas Law

Another factor to be considered in predicting a substance's state of matter is pressure. While pressure does not affect the energy level of an atom or molecule, forcing them closer together will increase the likelihood that molecules will bond with one another. Likewise, decreasing the pressure will allow for these small particles to move away from one another with less energy. So while pressure does not change the energy level of the small particles, it does change the energy level required by the particles to move away from one another. This relationship between pressure, volume, and temperature is explained by the *Ideal Gas Law*.

Ideal Gas Law Equation:

$PV = nRT$ 	P = pressure
	V = volume
	n = the amount of particles (measured in moles)
	R = the gas constant
	T = temperature

The equation above shows that volume and pressure are both proportionally related to temperature. This means that if the temperature is doubled, either the volume or the pressure will double as well (as long as the other remains consistent). If the temperature is decreased by a third, then either the volume or the pressure will decrease by a third as well (as long as the other remains consistent).

There is also a notable relationship between volume and pressure. These variables are inversely proportional to each other. If one is doubled, the other will decrease to half of its original quantity. If either volume or pressure is tripled, the other variable will decrease to one-third of its original quantity. This is assuming that the temperature is remaining constant.

Changing any of the three variables (temperature, volume, or pressure) can cause a phase change to occur. Gases can condense to a liquid or go through deposition to a solid. Liquids can freeze to a solid. Likewise, a solid can melt to a liquid or go through sublimation to a gas. Liquids can evaporate or vaporize into a gas. While gas can also change to plasma, this is commonly done through an extreme increase in temperature.

Phase change diagram

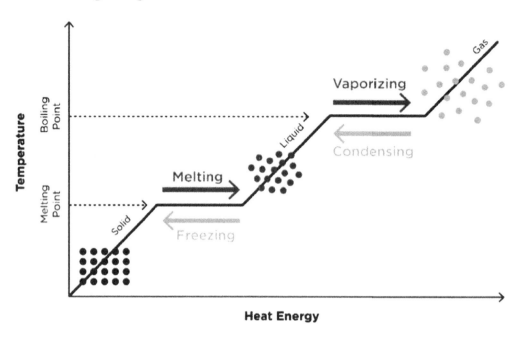

Heat and Temperature

An increase in heat is generally shown by an increase in temperature. This is true as long as the substance being measured is not going through a phase change. Adding heat during a phase change does not cause an increase in temperature. Temperature measures kinetic energy, but the energy in a phase change is potential energy. There is no temperature increase from the time the phase change starts until the phase change is complete. The term for the amount of heat required to complete phase changes is *latent heat*. The *latent heat of fusion* is the amount of heat required to change a substance from a solid to a liquid. The *latent heat of vaporization* is the amount of heat required for a substance to change from a liquid to a gas. Every substance has a specific latent heat of fusion and specific latent heat of vaporization.

Phase diagrams are useful tools that show how each state of matter, or phase, is determined by both temperature and pressure. Areas on these charts, called *heating curves,* give a visual representation of each phase and show how changing temperature, pressure, or both may alter a substance's physical state.

Chemical Bonding and Naming

<u>Reading Chemical Formulas</u>
Chemical formulas are made up of chemical symbols from the periodic table and numbers written as a subscript to the right of each symbol to show the ratio of how the symbols bond together. If a symbol does not have a subscript, the number 1 is implied. When parentheses are used, the subscript refers to everything inside the parentheses.

Examples:
> NaCl is made up of one sodium atom for every one chlorine atom.
> H_2O is made up of two hydrogen atoms for every one oxygen atom.
> $Al_2(SO_4)_3$ is made up of two atoms of aluminum, three atoms of sulfur, and 12 atoms of oxygen.

<u>Ionic Bonding</u>
Ionic bonding, which involves the transfer of electrons from one atom to another, is one way that elements bond together to create compounds. All chemical compounds are held together by the attraction of one atom to another. This attraction can be best visualized by thinking of magnets that are attracted to one another. Ionic bonding can involve a very strong attraction and results in atoms bonding in a crystal-like structure.

Ionic bonds require atoms to either gain or lose valence electrons (relative to the neutral state with an equal number of protons and electrons) and become charged ions. Whether an atom gains or loses electrons, and the number of electrons that it will gain or lose, can be predicted by the column, or group, that it belongs to on the periodic table of elements. If the atom loses an electron it becomes a *cation* with a positive charge. The charge is equal to the number of electrons it loses. Most cations have charges of 1+, 2+ or 3+. If an atom gains electrons, it becomes an *anion* that is negatively charged. Most anions have a charge of 1-, 2-, or 3-.

When an atom becomes an ion, its name changes. Positive ions, or cations, keep their name, but the word "-ion" is added as well. For example, when a sodium atom loses an electron, it becomes a sodium ion. When an atom gains electrons and becomes a negative ion, or anion, its name changes to end in "-ide". For instance, fluorine becomes fluoride.

An ionic compound is the result of a cation and an anion bonding through the attraction of their opposite charges. Ionic compounds are named by stating the name of the cation ion and then the anion. For instance, when a sodium ion and a fluorine ion bond, the compound is called sodium fluoride. The symbol for sodium fluoride is NaF. When writing the chemical symbol for an ionically bonded compound, the symbol for the cation is always written first, followed by the symbol for the anion.

Ions must have equal but opposite charges when they bond. In the case of sodium fluoride, a sodium ion has a 1+ charge and a fluoride ion has a 1- charge, so they bond easily in a one-to-one ratio. For a magnesium ion with a 2+ charge to bond with a fluoride ion with a 1- charge, two fluoride ions must bond to every magnesium ion to make the charges equal and opposite (2+ bonding with 2-). The formula for magnesium fluoride is written MgF_2 to show that two fluoride ions bond with one magnesium ion.

Covalent Bonding
Another common type of bond between atoms is a *covalent bond,* which involves the sharing of valence electrons between atoms. Covalent bonding is not as easy to predict as ionic bonding, and there are many more ways elements can bond with one another than they can in ionic bonding. For this reason, it is best to write the chemical name from the formula or to create the formula from the name. A basic rule for naming a compound resulting from a covalent bond is to add a prefix to the front of an element's name to indicate the number of atoms of that element that makes up the compound. The prefixes include mono-, di-, tri-, tetra-, penta-, hexa-, hepta-, octa, nona, deca-, and so on. If the first element named has only one atom in the compound, the "mono-" is dropped. Another naming rule for covalent compounds is to end the last element's name in "-ide". For instance, CO_2, which is made up of one carbon atom and two oxygen atoms, is called carbon dioxide. CCl_4 is named carbon tetrachloride and made up of one carbon atom and four chlorine atoms.

Lewis Dot Diagrams
Lewis dot diagrams, named for Gilbert Lewis, create a visual model of bonding electrons on single elements or in a compound to represent the way atoms in an element or compound are bonded together. For a single atom of an element, Lewis dot diagrams show valence electrons surrounding the element's symbol. The symbol represents the protons, neutrons, and electrons of the inner electron shells. The dots around the outside represent the electrons in the outer shell, or valence electrons. Up to eight valence electrons can be represented by imagining an invisible square drawn around the symbol and placing two electrons on each side, as demonstrated by the Lewis dot diagram for Neon below:

For most elements with fewer than eight valence electrons, dots must be placed individually on each side before any side can receive two dots:

Lewis dot structures for compounds use lines between structures to indicate a shared bond of two electrons, and dots around the outside of elements to indicate the location of other outer electrons:

The Mole

While chemical formulas are fantastic for understanding the ratio of atoms that form a specific compound, scientists do not work with atoms in such small groups. When working with chemicals, scientists use mass to measure the amount of chemicals to use. Every element has a different mass and atoms are too small to work with individually, so a different type of calculation must be used to determine the mass of each element needed in order to have the correct number of particles in a reaction. *Stoichiometry* is the study of the quantities used in chemical reactions and involves the calculation that converts between the mass of an element or compound and the number of particles that it contains. In stoichiometry, a special conversion number is used. 6.022×10^{23}, also known as Avogadro's number, is the converting number between mass and particles. This number is also called the *mole*. Other common converting numbers are a dozen, pair, ream, and gross. All of these terms represent a specific number of an item. None of them have their own units. The mole, 6.022×10^{23}, is no exception.

Molar Mass

A *molar mass* is the mass of one mole of any substance. The masses listed on the periodic table are equal to the amount of grams in one mole of that substance. While those masses can have the units of amu when referring to one atom, they can also have the units of grams per mole (g/mol) when referring to 6.022×10^{23} atoms of that element.

To find the molar mass of a compound, start by multiplying the mass of each element in the compound (as listed on the periodic table) by its subscript from the formula. Then add all of the totals together to determine the total mass in grams per mole (g/mol) of that substance.

Example: To find the molar mass of glucose, $C_6H_{12}O_6$, the mass of carbon (12.01 g/mol) is multiplied by 6 (its subscript in the formula) to get 72.06 g/mol. The mass of hydrogen (1.01 g/mol) is multiplied by 12 to get 12.12 g/mol. The mass of oxygen (16.00 g/mol) is multiplied by 6 to get 96.00 g/mol. These three totals (72.06 + 12.12 + 96.00) are added together to get the total molar mass of glucose, 180.18 g/mol.

Percent Composition

Percent composition is a formula used to determine the percent by mass of each element making up a compound. To determine the percent composition of a compound, divide the molar mass of each element by the molar mass of the entire compound and multiply the answer by 100.

For example, here is how we determine the percent composition of water, H_2O:

Molar mass of hydrogen = 1.01 g/mol × 2 mol hydrogen = 2.02 g H
Molar mass of oxygen = 16.00 g/mol O × 1 mol O = 16.00 g O
Molar mass of H_2O = (1.01 g/mol H × 2 mol H) + (16.00 g/mol O × 1 mol O) = 18.02 g H_2O

Percent composition of hydrogen = $\frac{2.02 \text{ g}}{18.02 \text{ g}}$ × 100 = 11.21%

Percent composition of oxygen = $\frac{16.00 \text{ g}}{18.02 \text{ g}}$ × 100 = 88.79%

A quick check to make sure the answers add up to 100 is an easy way to make sure there are no mistakes.

Chemical Reactions

Chemical Equations

Chemical equations use symbols to represent the details of chemical reactions. These equations include which elements or compounds are present at the beginning of a reaction (*reactants*) and which elements or compounds are the result of the reaction (*products*). The format for writing a chemical equation generally includes a listing of reactants separated by the plus sign ("+") followed by an arrow, and then a listing of the products separated by the plus sign ("+"). If a catalyst or energy is used, it is written above the arrow.

Example:

$$2H_2 + O_2 \rightarrow 2H_2O$$
(reactants)   (products)

Notice the coefficient in front of H_2 on the reactant side of the equation as well as in front of H_2O on the product side. These coefficients are placed as a way to balance the equation. Because of the law of conservation of matter, the number of particles that start an equation must be equal to the number of particles (and of the same kind) that are produced by the equation. As elements and compounds exist in nature in fixed ratios, their subscripts cannot be changed in a chemical equation. This can result in an equation that does not obey the law of conservation of matter. Coefficients are used to balance the equation on both sides of the arrow so that the law may be obeyed.

Types of Chemical Reactions

Chemical reactions can be categorized into several different types of reactions. A *synthesis*, or *combination*, reaction involves smaller reactants being combined into larger products. The reaction of lithium metal combining with fluorine gas to create the compound lithium fluoride is an example of a synthesis reaction:

$$2Li + F_2 \rightarrow 2LiF$$

Decomposition reactions are reactions where larger reactants break down into smaller products. For example, calcium carbonate breaks down when exposed to heat to create calcium oxide and carbon dioxide, as shown below:

$$CaCO_3 \rightarrow CaO + CO_2$$

Combustion reactions require pure oxygen as a reactant and release a large amount of energy as heat or light. The burning of propane is a combustion reaction:

$$C_3H_8 + 5O_2 \rightarrow 3CO_2 + 4H_2O$$

Single Replacement reactions, or *single displacement* reactions, involve a single element replacing one component in a compound. In the following reaction, hydrochloric acid turns into nickel chloride and releases hydrogen gas, as chlorine is more attracted to nickel than it is to hydrogen:

$$Ni + 2HCl \rightarrow NiCl_2 + H_2$$

Double Replacement reactions, or *double displacement* reactions, have two compounds as reactants. During the reaction, these compounds separate into their component ions and then rebond as new compounds. The end result is a swap of ions between the two compounds. For instance, when solutions of silver nitrate and potassium chloride are mixed, the resulting products are silver chloride and potassium nitrate:

$$AgNO_3 + KCl \rightarrow AgCl + KNO_3$$

Organic reactions are any reactions that have carbon and hydrogen as components. *Organic chemistry* is an entire branch of chemistry dedicated to studying organic reactions. Cellular respiration is an organic reaction that can be represented by the equation below:

$$C_6H_{12}O_6 + 6O_2 \rightarrow 6H_2O + 6CO_2$$

Oxidation/reduction reactions are reactions that involve the loss of electrons from one component (oxidation) and the gain of electrons by another component (reduction). These reactions can also be referred to as *redox* reactions or *half reactions.* In the following reaction, the oxidation of iron as it comes in contact with nickel nitrate is accompanied by the reduction of nickel as iron nitrate and nickel are formed:

$$Fe + Ni(NO_3)_2 \rightarrow Fe(NO_3)_2 + Ni$$

Factors That Affect the Rate of Reaction

The time that is required for different elements and compounds to completely react with one another can vary. The specific elements and compounds in the reaction can be the major determining factor in the rate of the reaction, but pressure, temperature, particle size, concentration, physical state, and the presence of catalysts or enzymes can also influence how quickly reactants interact with one another. The rate of reaction ultimately depends on the ability of atoms and molecules to bump into one another, so that they can react. Increasing pressure will decrease the volume, which limits the space that the particles have. This will encourage more interactions in a shorter amount of time. Increasing the energy in the molecules with the addition of a catalyst or heat will increase the speed at which they are moving. A smaller particle size increases the amount of available surface area. Particles in fluids not only move more quickly than particles in a solid, but they are much more mobile in their ability to flow past one another.

Beyond increasing the rate of reaction, these same variables can be used to stimulate further reactions once the initial reaction has stopped due to equilibrium being achieved. *Le Chatelier's principle* states that equilibrium of the reaction will shift in response to a change in pressure, temperature, or concentration. Using knowledge of equilibrium and molecular motion allows chemists to use outside forces to encourage a different rate of reaction in the planning of their experiments.

Energy of Chemical Reactions

Energy is either released or absorbed in a chemical reaction. When energy exits the system, the reaction is *exothermic*. When energy enters the system, the reaction is *endothermic*. Exothermic reactions will make the container it is reacting in feel warm or even hot. Endothermic reactions will make the container feel cool or cold to the touch.

Acid-Base Chemistry

Acids and Bases

Acids and bases are usually defined by the presence of hydrogen ions (H+) or hydroxide (OH-) in a solution. The *pH scale* is a logarithmic (base ten) scale based on the concentration of H^+ in a solution, with a pH value of 7 being considered neutral. With equal parts H^+ and OH^-, water is neutral and has a pH value of 7. As pH values increase from 7 toward the upper limit of 14, solutions become more basic. Similarly, solutions become more acidic as their values decrease toward the bottom of the scale, which is 0.

Characteristics of a solution that indicate the presence of a base include feeling slimy to the touch, turning red litmus paper blue, turning phenolphthalein pink, and tasting bitter. Characteristics of a solution that indicate the presence of an acid include producing bubbles when introduced to a metal, turning blue litmus paper red, turning phenolphthalein colorless, and tasting sour.

pH scale

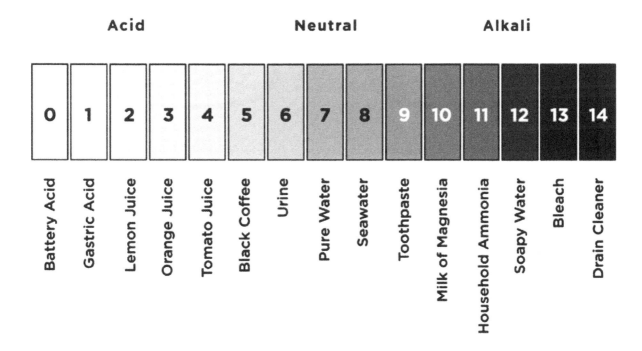

Solutions and Solubility

Different Types of Solutions

Solutions are homogeneous mixtures containing more than one substance. These mixtures have a uniform composition throughout and are composed of a *solute* and a *solvent*. The solvent is usually the component of the solution present in a larger quantity. The solute is dissolved in the solvent and present in a smaller quantity. In a solution of salt water, salt is the solute, and water is the solvent. Water is used so frequently as a solvent in a solution that it is referred to as the "universal solvent." When a solute is added to a solvent, it begins to dissolve, and the particles move freely throughout the solution. When solute particles are present in small amounts, a *diluted* solution exists. As more solute is added, the solution can become a *concentrated* solution.

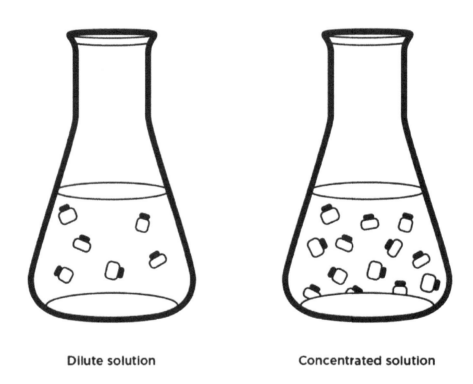

Dilute solution **Concentrated solution**

As a solution increases in concentration, solute particles begin to encounter one another and crystalize back into solid form. In a dilute solution, more solute particles are dissolving than are crystalizing back into solid form. Increasing the concentration of solute particles increases the amount of *crystallization*. Eventually, as the concentration increases to a level where the rate of crystallization equals the rate of particles dissolving, the *saturation point* is reached, and the solution is considered *saturated*.

Saturated solutions cannot dissolve more solute unless forced by manipulating the temperature or another variable of the solution. When this is done, it is possible to create a *supersaturated* solution, but the variables must remain constant or crystallization can occur. A good example of this is the creation of rock candy. Using water as a solvent and sugar as a solute, a supersaturated solution can be created by increasing the temperature of the solution in order to dissolve more sugar. Putting a stirring stick, straw, or other item in the solution and allowing it to cool will result in the crystallization of sugar, producing a sweet treat! Adding food coloring or artificial flavors makes this experiment in supersaturated solutions even more enjoyable.

When working in a lab, the descriptions of *dilute, concentrated, saturated, supersaturated, saturation point,* and *crystallization* are not sufficient, as they are qualitative observations. The accuracy of a quantitative measurement is needed when working with acids, bases, or reactive chemical solutions. Concentration, or *molarity*, is used to quantify the exact amount of solute dissolved in a solution. Molarity is equal to moles of solute per liter of solution.

The formula for determining molarity is:

$c = n/V$

c = molarity (moles/liter)
n = moles of solute
V = volume (liters)

Example problem when solving for molarity (M): What is the molarity of a solution made by dissolving 45.0 grams of Na_2SO_4 in enough water to make 100 mL of solution?

Step 1: Find the molar mass of Na_2SO_4:

Na → 2 x 23.0 grams/mole = 46.0 grams/mole Na
S → 1 x 32.1 grams/mole = 32.1 grams/mole S
O → 4 x 16.0 grams/mole = 64.0 grams/mole O

46.0 grams/mole Na + 32.1 grams/mole S + 64.0 grams/mole O
= 142.1 grams/mole Na_2SO_4

Step 2: Use dimensional analysis and the molar mass of Na_2SO_4 as a conversion rate to determine the amount of moles of Na_2SO_4 in 45.0 g Na_2SO_4:

(45.0 g Na_2SO_4) x (1 mole Na_2SO_4/142.1 grams Na_2SO_4) = 0.317 moles Na_2SO_4

Step 3: Divide the moles of solute (Na_2SO_4) by the solution in L:

0.317 moles Na_2SO_4/0.100 L = 3.17 M Na_2SO_4

Example problem when solving for mass solute in grams: How many grams of Na_2SO_4 are required to make 0.400 L of 2.00 M Na_2SO_4?

Step 1: Use dimensional analysis to convert 0.400 L into moles. (Remember that M=mole/liter gives us the ability to use molarity as a conversion rate):

(0.400 L solution) x (2.00 mole Na_2SO_4/1.00 L solution) = 0.800 mole Na_2SO_4

Step 2: Use dimensional analysis and the molar mass of Na_2SO_4 to convert moles of Na_2SO_4 into grams of Na_2SO_4

(0.800 mole Na_2SO_4) x (142.1 grams Na_2SO_4/1 mol Na_2SO_4) = 114 grams Na_2SO_4

Reactions Between Solutions

Solutions are a useful way to mix solutes that don't mix readily in their solid form. Once dissolved, solutes have already broken the intermolecular forces holding them together. When solutions are mixed together, the constituent particles of the solutes are free to bond with other particles that they come in contact with. When mixing an acidic solution with a basic solution, a very interesting result occurs. The H^+ ions from the acid and the OH^- ions from the base come together to form water and a new ion pair (known as a salt, but it is not necessarily table salt). It is interesting that the result of mixing two dangerous solutions together is simply salt water. Depending on what ions combine, a *precipitate*, or solid form of a salt, may be produced.

Polar vs Nonpolar Solvents

Although water is considered "the universal solvent" and is frequently used when making solutions, not all solutes will dissolve in water. A water molecule has a partial positive charge on one end, and a partial negative charge on the other end. This is due to an uneven sharing of electrons in the covalent bonds between hydrogen and water. The measure of these charges is called *electronegativity*. Particles with different electronegativities, like water, are considered *polar*. Particles with the same electronegativities are considered to be *nonpolar* particles. Both solvents and solutes can be polar or nonpolar. Scientists use the phrase "like dissolves like" to remember that polar solutes dissolve best in polar solvents and nonpolar solutes dissolve best in nonpolar solvents. An example of a nonpolar solvent is benzene.

Variables Affecting Solubility Rates

Manipulating different variables can create an environment suitable for achieving supersaturation. These same variables can also be manipulated to change the rate of solubility, or how quickly a solute dissolves in a solution. It is useful to consider the motion of the particles of the solute when considering ways to manipulate solubility. The ability of the particles to overcome intermolecular forces can be increased by giving them more energy in the form of heat. Increasing temperature will increase the rate that they dissolve. Likewise, increasing pressure will increase the rate of solubility, too. Adding mechanical energy to the system by simply stirring the solution will also increase the rate of solubility. The more agitation (stirring) that is added, the faster the solute will dissolve. Decreasing particle size will increase the surface area of the solute and give more particles the ability to dissolve more quickly. Lastly, manipulating the concentration of solute or solvent will change the solubility rate. Increasing the volume of solvent will decrease the likelihood of solute particles running into each other and crystallizing. Many of these variables can be recognized in the *ideal gas law,* $PV=nRT$.

Solubility Curves
The amount of solute that can dissolve in a solution will increase with increasing temperatures. This is not usually a direct relationship. Every solute increases at a different rate with increased temperature. To understand this relationship for a particular solute, its *solubility curve* can be referenced. A solubility curve graphs the relationship between the temperature on the x-axis and the mass in grams of solute on the y-axis. Any coordinates above the curve represent a temperature and mass combination that would be *unstable* or *supersaturated*. Coordinates below the curve represent a combination of temperature and mass that would create an *unsaturated* solution. If a temperature and mass combination fall on the solubility curve, the solution is *saturated*.

Freezing Point Depression
Water has a freezing point of 0 °C. When the temperature outside drops below this, any water on the streets or sidewalks freezes and creates icy conditions that can cause accidents for cars and pedestrians. There are also fluids in our cars that can freeze and cause problems for winter drivers. Fortunately, manipulating the freezing point of water and other fluids can be done by adding solutes and creating a solution. Adding a solute of salt to water lowers the freezing point, creating safer conditions when the weather is a few degrees below the freezing point of water, but above the freezing point of the salt water solution. Mixing ethylene glycol in radiator fluid does the same thing for a car. Using a solute in order to lower the freezing point of a solvent is called *freezing point depression.*

Before You Start the Practice Questions

On the next page, you will begin a Physical Science practice test. Set a timer for 10 minutes before you start this practice test. Giving yourself 10 minutes will give you an authentic feel for how long you have to finish this portion of the AFOQT. Like the official test, this practice test has exactly 20 items. Set your timer, turn the page, and begin.

Physical Science Practice Test Questions

(10 Minutes)

1. Which of the following best describes a bucket of gravel and water?
    a. A homogeneous mixture
    b. A heterogeneous mixture
    c. A solution
    d. An ionic compound
    e. A covalent compound

2. Where does a roller coaster car have the most kinetic energy?
    a. At the top of a hill
    b. As it is descending a hill
    c. At the bottom of a hill
    d. As it is ascending a hill
    e. As it comes to a complete stop at the end of the track

3. Which of the following is an example of a phase change?
    a. Leaves changing colors
    b. Wood burning in a fire
    c. Bread molding
    d. Iron rusting
    e. Water vaporizing

4. Which of the following is best explained by the concept of *specific heat capacity*?
    a. A hotel keeps its pool at exactly 85°F because it is the ideal temperature enjoyed by a majority of its patrons.
    b. 145°F is the internal temperature needed to safely cook a pork loin.
    c. It takes the same amount of time to boil 1 cup of water as it does 1 gallon of water when the same amount of heat and pressure are applied.
    d. On a very hot day, ocean water can be refreshing even when the sand on the beach burns your feet.
    e. Putting salt on an icy sidewalk in winter melts the ice by lowering the freezing point of water.

5. What is the best description of the relationship between temperature and energy as ice melts?
    a. The temperature remains at the melting point as the potential energy of the system increases.
    b. The temperature slowly increases as the potential energy of the system increases.
    c. The temperature remains the same as the melting point as the kinetic energy of the system increases.
    d. The temperature slowly increases as the kinetic energy of the system increases.
    e. The temperature increases, but the energy of the system does not change.

6. What is true about an atom of an element with an atomic number of 92?
    a. It has a mass of 92 grams per mole.
    b. It has a mass of 184 atomic mass units.
    c. It has 92 electrons surrounding the nucleus.
    d. It has 92 neutrons in the nucleus.
    e. It has 92 protons in the nucleus.

7. How many atoms of each element are in the ionic compound calcium phosphate, $Ca_3(PO_4)_2$?
    a. 3 atoms of calcium, 2 atoms of phosphorus, 6 atoms of oxygen
    b. 3 atoms of calcium, 1 atom of phosphorus, 4 atoms of oxygen
    c. 3 atoms of calcium, 2 atoms of phosphorus, 8 atoms of oxygen
    d. 1 atoms of calcium, 1 atom of phosphorus, 1 atom of oxygen
    e. 3 atoms of calcium, 2 atoms of potassium, 6 atoms of oxygen

8. Which of the following is the correct name for the covalently bonded molecule PH_3?
    a. Trihydrogen phosphide
    b. Monophosphorus trihydride
    c. Trihydrogen monophosphide
    d. Phosphorus trihydride
    e. Phosphorus hydride

9. How many moles of water are contained in 36.04 grams of water?
    a. 1 mole of water
    b. 2 moles of water
    c. 3 moles of water
    d. 6.022 x 10^{23} moles of water
    e. 2.32 moles of water

10. How many molecules are contained in a sample of 100.0 grams of carbon dioxide?
    a. 6.022 x 10^{23}
    b. 2.150 x 10^{24}
    c. 1.368 x 10^{23}
    d. 2.272
    e. 1.368 x 10^{24}

11. Which phase change is an endothermic process?
    a. Freezing
    b. Condensing
    c. Deposition
    d. Melting
    e. Solidifying

12. Which of the following can affect the rate of a chemical reaction?
    a. Temperature
    b. Presence of a catalyst
    c. Pressure
    d. Concentration of reactants
    e. All of these answers are correct

13. Water is not considered an acid or a base when it is alone. Which of the following is true about a neutral substance like water?
    a. It has a pH of 7.
    b. It has a pH of 14.
    c. It has a pH of 0.
    d. It has more H^+ than OH^-.
    e. It has more OH^- than H^+.

14. A nonpolar solute will dissolve best in a solvent that
    a. is nonpolar.
    b. is polar.
    c. has large electronegativity differences on the ends of its molecules.
    d. is water.
    e. is ionically bonded.

15. According to Newton's Laws of Motion, which of the following is true about punching a brick wall?
    a. A lot of force is exerted on the wall, but none on the fist.
    b. A lot of force is exerted on the fist, but none on the wall.
    c. More force is exerted on the wall than on the fist.
    d. More force is exerted on the fist than on the wall.
    e. The force exerted on the fist is equal to the force exerted on the wall.

16. Which of the following is true about centripetal forces?
    a. They are fictitious forces.
    b. They are an outward pulling force.
    c. They pull items out of a spin and onto a path tangential to the path of their previous rotation.
    d. They act in the opposite direction as a centrifugal force.
    e. They are an inward pulling force.

17. Which sentence is incorrect about simple machines?
    a. Machines can be used to change the direction of the required force.
    b. Machines can be used to change the amount of the required force.
    c. Machines can change mechanical advantage by increasing the output force.
    d. Machines can change the amount of work that must be done to lift something.
    e. Machines can lessen the amount of force required to do work by increasing the distance over which it is done.

18. Which of the following encompasses the smallest section of the electromagnetic spectrum?
    a. Visible Light
    b. Ultraviolet Light
    c. Microwaves
    d. Gamma Waves
    e. Infrared Waves

19. Which of the following best explains why a red shirt looks red?
    a. The shirt's chemicals are red.
    b. The shirt is absorbing red light.
    c. The shirt is absorbing every color of light except red.
    d. The person wearing the shirt is moving closer to you.
    e. The light hitting the shirt is predominantly red.

20. Which of the following is incorrect about sound waves?
    a. The magnitude of amplitude in a sound wave is related to loudness.
    b. Sound waves are transverse waves.
    c. Sound waves have areas of both compression and rarefaction.
    d. The pitch of a note is related to the frequency of a sound wave.
    e. Sound waves require a medium to travel.

Answer Guide to Physical Science Practice Test

1. **B**. Heterogenous mixtures are not evenly mixed throughout. Choice A is incorrect because homogenous mixtures are mixtures that are uniform throughout and not easily filtered. Choice C is a specific type of homogenous mixture. Choices D and E describe something that has chemically bonded. The water and gravel have physically mixed, but not chemically bonded.

2. **C**. The sum of potential and kinetic energy will be the same for the roller coaster car no matter where it is on the track, so the place where it has the most kinetic energy will be the place where it has the least amount of potential energy. Potential energy is dependent on how high off the ground the car is positioned. Choice A is the highest point, so it would have the highest potential energy and lowest kinetic energy. Choices B and D still have height greater than choice C, so they have more potential energy and thereby less kinetic energy. The roller coaster car has no motion in choice E. Kinetic energy is directly proportional to velocity. If there is no velocity, there is no kinetic energy.

3. **E**. Water vaporizing is a phase change. The chemical composition of water is not changed when it changes from a liquid to a gas. It can easily be turned back into a liquid. Choices A, B, C, and D all represent chemical changes. The chemical compositions of the objects in these choices are all altered. Reversing any of these changes would be impossible.

4. **D**. Water has a much higher specific heat capacity than sand, so while both are subjected to the same amount of energy from the sun, the water will change temperature at a much slower rate. Choices A and B are just temperatures that are preferable for specific conditions. They have nothing to do with specific heat capacity. Choice C is not a true statement; because of the difference in mass, it would take longer to heat a larger amount of water. This can be shown with the formula used with specific heat: $q = sm\Delta t$. Choice E describes the concept of freezing point depression, not specific heat capacity.

5. **A**. During any phase change, the temperature remains the same until the phase change is complete. Because temperature measures average molecular kinetic energy and it is not changing, there is no change in kinetic energy. Due to the law of conservation of energy, energy is being added to a system when it is heated. If there is no temperature change, the energy added to the system has to be potential.

6. **E.** The atomic number is always equal to the number of protons in the nucleus of an atom of that element. Choice A describes an atomic mass, which is a weighted average mass of all the isotopes of an element's atoms. Choice B describes an atomic mass number, which is the sum of the number of protons and the number of neutrons in a single atom's nucleus. Choice C is not an accurate answer; while the number of electrons frequently equals the number of protons in an electrically neutral atom, this is not always the case. The electrons can change in number, and the atomic number will stay the same. However, if the protons in the nucleus of an atom were to change, the atomic number would also change. Choice D is not an accurate answer; the number of neutrons in an atom's nucleus can certainly be equal to the number of protons in the nucleus, but there are many isotopes of the same atom where this is not the case. Larger atoms are much less likely to have an equal number of protons and neutrons. The number of protons determines the atomic number. The number of neutrons does not.

7. **C.** The subscripts designate how many atoms of each element are in the compound. The phosphate group is made up of 1 phosphorus and 4 oxygen atoms. With a subscript of 2 outside the parentheses, there are two complete phosphate groups, which makes a total of 2 phosphorus atoms and 8 oxygen atoms. There are 3 calcium atoms. Choice A has too few oxygen atoms. Choice B has too few phosphorus and too few oxygen atoms. Choice D has too few of every element. Choice E lists potassium instead of phosphorus and has too few oxygen atoms.

8. **D.** In covalent bonding, the names are in the same order as the written formula, with the second element ending in "-ide." Prefixes are used to denote the number of each atom present. The "mono-" is dropped off of the first element when it only contains 1 atom. Choice A has the elements in the wrong order. If the elements were listed this way, there would be the prefix "mono-" on the second element. Choice B doesn't drop the prefix "mono-" from the first element. Choice C is also in the wrong order. Choice E has no prefixes.

9. **B.** Water is composed of 2 hydrogen atoms and 1 oxygen atom. According to the periodic table, the molar mass of hydrogen is 1.01 grams/mole. Because there are 2 atoms of hydrogen, 1 mole of water has 2.02 grams of hydrogen. The molar mass of oxygen, according to the periodic table, is 16.00 grams/mole. Because there is one oxygen atom, a mole of water has 16.00 grams of oxygen. Adding the hydrogen mass and the oxygen mass together gives us a molar mass for water of 18.02 grams/mole. Because molar mass is the mass in grams for every one mole, dividing the mass in the problem (36.04 grams) by the molar mass of water gives us the number of moles. 36.04/18.02 = 2 moles of water. Choice A would only

weigh 18.01 grams. Choice C would weigh 54.06 grams. Choice D is Avogadro's number and is the number of particles per mole of any given item. Choice E would weigh 41.81 grams.

10. **E.** The formula for carbon dioxide is CO_2. The molar mass would be equal to the sum of the molar masses of one carbon atom (12.011 g/mol) and two oxygen atoms (32.000 g/mol), or 44.011 g/mol. To determine the number of moles there are, the mass of 100.0 grams should be divided by the molar mass of 44.011 g/mol. This yields 2.272 moles of CO_2. One mole of anything is equal to 6.022×10^{23} of that item. Multiplying 2.272 by 6.022×10^{23} gives the correct answer of 1.368×10^{24} molecules of CO_2. Choice A is the number of particles in one mole. Choice B would be the answer if there were only one oxygen in the formula for CO_2. Choice C is incorrect by a power of 10. Choice D is the number of moles, not the number of molecules.

11. **D.** An endothermic process is one that brings energy into the system from its surroundings. Melting a solid into a liquid requires giving the molecules more energy to overcome intermolecular attractions and allow them to move around. Choices A, B, and C are all processes where the molecules are releasing energy and slowing down. They are exothermic processes. Choice E is technically not a phase change but could be considered another name for freezing.

12. **E.** Temperature, concentration of reactants, the presence of a catalyst or enzyme, and pressure are all ways to change the rate of a chemical reaction.

13. **A.** A neutral solution has a pH of 7 with equal amounts of H^+ and OH^-. Choices B and E would be considered basic, having more OH^- than H^+ and a pH of greater than 7. Choices C and D would be considered acidic, having more H^+ than OH^- and a pH of less than 7.

14. **A.** Chemists use the saying "like dissolves like" when determining the most effective solvent for dissolving a solute. A nonpolar solute would need a nonpolar solvent. Choice B is incorrect as it is not "like" the nonpolar solute. Choice C is incorrect because electronegativity differences are what cause polarity. Choice D is incorrect; ionic bonds create the most polar compounds. Choice E is incorrect; water is known for being a polar solvent.

15. **E.** Forces always occur in pairs that are equal in magnitude but opposite in direction. If choice A were correct, there would be no damage done to the fist, but punching a wall causes a lot of damage to a fist. There is a direct push (force) on the wall with the fist, so choice B is incorrect. The forces on the wall and the fist

are equal, but more damage is done to the hand than is done to the wall because the mass of the wall is so much greater than that of the hand. Therefore both choices C and D are incorrect.

16. **E.** A centripetal force is any force that keeps an object in its rotational path and is directed toward the center or axis. Choice A is not correct; centrifugal forces are nicknamed "fictitious forces." Choice B is incorrect; centripetal forces are directed toward the center. Choice C is incorrect; centripetal forces are what keep a spinning object in its rotational path. Choice D is incorrect; a centrifugal force is not part of a force interaction. It is simply the lack of centripetal force.

17. **D.** D is the only incorrect statement because machines don't change the amount of work that must be put into something to lift it. Work is the product of both distance and force. Machines can work by lengthening the distance, so the force required is less to do the same amount of work. They can also create a mechanical advantage by changing the ratio of input force to output force, or just change the direction of the needed force.

18. **A.** Visible light is a very small portion of the electromagnetic spectrum. Choices B, C, D, and E all encompass much larger sections.

19. **C.** The color you see depends on the wavelength of light being reflected back to you. Red is being reflected from the shirt while all other colors of light are being absorbed. Choice A is not technically correct because the chemicals themselves do not have a color. If they were to appear red, it would be for the same reason that makes choice C correct. Choice B is not correct because the wavelengths of light that get absorbed are not visible. Choice D could technically be correct if it were a star showing red shift, but we don't experience a visible Doppler effect for slow moving objects here on Earth. Choice E wouldn't be correct unless the shirt were being viewed in a limited location where light is filtered to only show red. Even in that situation, the shirt itself would be no redder than the rest of the environment.

20. **B.** Sound waves are longitudinal waves. Choices A, C, D, and E are all correct statements. Because sound waves are longitudinal, they do have areas of compression and rarefaction. Sound waves are mechanical waves and not electromagnetic waves, so they require a medium to travel.

TABLE READING

In this section of the AFOQT, you will be tested to see if you can accurately read a table of numbers. If you look below, you will see a chart of numbers. The horizontal set of numbers is the x-axis (left to right). The vertical set of numbers is the y-axis (top to bottom). The far left of the x-values reads "-3." The top of the y-values reads "+3." The x-values are considered the columns of data. The y-values are considered the rows. If you are given an x-value and y-value, you will find the number at that coordinate. You will choose the correct answer from five options.

Analyzing the table below will help you to better understand the types of questions you will encounter during this part of the test. For instance, if you are given the coordinates (-3, 2), that would mean the x-value is -3, and the y-value is +2. If you look at column -3 and go down to row +2, you will see that the number is 22.

You will notice that each question (1 through 5) has five options (a through e).

X-Values

		-3	-2	-1	0	+1	+2	+3
	+3	20	22	24	23	21	25	26
	+2	22	19	24	28	27	28	21
	+1	23	33	30	29	31	18	22
Y-Values	0	23	23	22	27	20	28	24
	-1	16	17	19	25	24	34	39
	-2	30	37	37	36	35	37	36
	-3	26	24	29	28	20	23	21

Try to accurately answer the problems below. Circle the correct answer for sample problems 1 through 5, using the table on the previous page:

	X	Y	A	B	C	D	E
1.	-2	+2	27	19	25	26	20
2.	0	-1	22	27	25	21	19
3.	-3	-1	16	23	29	18	21
4.	+1	+2	31	23	19	27	18
5.	-2	+1	39	35	22	30	33

Answers to Sample Problems:

1. **B**
2. **C**
3. **A**
4. **D**
5. **E**

Explanations for Sample Problems:

1. In question 1, you are given the coordinates -2 and +2. The x-value (left to right) is -2, and the y-value (top to bottom) is +2. The number in that box is 19. (option **B**).
2. In question 2, you are given the x-value 0 and y-value -1. Find column 0 and row -1. They intersect at box 25. (option **C**).
3. Question 3 gives you x-value -3 and y-value -1. Column -3 is on the far left and row -1 is five boxes down. The answer is 16. (option **A**).
4. Question 4 gives you x-value +1 and y-value +2. The number in that box is 27. (option **D**).
5. Coordinates (-2, 1) mean x-value is -2 and y-value is +1. The answer is 33. (option **E**).

Test Tips

- Start with the x-axis and find the correct column (-3 through +3), then move down to the correct row on the y-axis to find the answer.
- Use your finger to move along the x and y columns and rows if it helps you find the answer.
- Use a piece of paper to cover the next problems and answer choices on the test to keep track of which problem you are on. Move the edge of the paper down each time you finish a problem to uncover the next problem and answer choices. NOTE: you are not permitted to use a straight edge on the data table itself, and you may be asked not to use a straight edge at all.
- Don't mix up x-values and y-values. It's easy to accidentally mistake the x-value for the y-value. If you have time left over on the test, you can double check your answers.
- Some people find this section of the test fairly easy and straightforward, so they take it lightly. Do not repeat this mistake. Even though you may find this section simple, take it seriously, and pay attention to the details. It's easy to make a mistake, so be diligent.
- **It is important to note that the official test can sometimes have much larger data tables than the ones included in this practice test. Come test day, be prepared to find x-values and y-values on larger tables.**
- Be quick, but don't be reckless. Pace yourself so you don't make a hurried mistake.
- Take the following practice test to make sure you are proficient in this section of the test.

Before You Start the Practice Questions

On the next page, you will begin a Table Reading practice test. Set a timer for 7 minutes before you start this practice test. Giving yourself 7 minutes will give you an authentic feel for how long you have to finish this portion of the AFOQT. Like the official test, this practice test has exactly 40 items. Set your timer, turn the page, and begin.

Table Reading Practice Test Questions
(7 minutes)

The next five questions are based on the following table:

	-3	-2	-1	0	+1	+2	+3
+3	21	22	23	26	27	25	24
+2	28	19	25	23	24	29	31
+1	30	29	28	25	23	22	21
0	27	32	31	20	22	24	28
-1	29	23	20	19	18	28	22
-2	25	27	22	21	20	19	23
-3	27	29	28	24	21	27	29

	X	Y	A	B	C	D	E
1.	-3	+2	22	28	21	31	23
2.	+3	-2	23	28	27	29	25
3.	-2	-1	28	24	23	25	22
4.	+2	+2	27	19	22	24	29
5.	0	+1	25	31	19	22	28

1. (-3, 2)
    a. 22
    b. 28
    c. 21
    d. 31
    e. 23

2. (3, -2)
    a. 23
    b. 28
    c. 27
    d. 29
    e. 25

3. (-2, -1)
    a. 28
    b. 24
    c. 23
    d. 25
    e. 22

4. (2, 2)
    a. 27
    b. 19
    c. 22
    d. 24
    e. 29

5. (0, 1)
    a. 25
    b. 31
    c. 19
    d. 22
    e. 28

The next five questions are based on the following table:

	-3	-2	-1	0	+1	+2	+3
+3	30	32	35	37	38	40	39
+2	32	33	34	36	37	32	31
+1	35	37	36	41	28	29	24
0	26	27	28	29	25	23	33
-1	34	33	39	40	37	36	35
-2	33	35	36	37	38	39	40
-3	29	32	31	35	36	37	38

	X	Y	A	B	C	D	E
6.	0	+3	32	26	33	37	35
7.	-2	-1	34	33	37	29	36
8.	-1	-1	37	39	28	29	30
9.	+2	+3	33	31	32	34	40
10.	0	+2	27	36	23	25	24

6. (0, 3)
    a. 32
    b. 26
    c. 33
    d. 37
    e. 35

7. (-2, -1)
    a. 34
    b. 33
    c. 37
    d. 29
    e. 36

8. (-1, -1)
    a. 37
    b. 39
    c. 28
    d. 29
    e. 30

9. (2, 3)
    a. 33
    b. 31
    c. 32
    d. 34
    e. 40

10. (0, 2)
    a. 27
    b. 36
    c. 23
    d. 25
    e. 24

The next five questions are based on the following table:

	-3	-2	-1	0	+1	+2	+3
+3	44	42	41	43	40	41	48
+2	47	46	45	44	42	40	49
+1	51	50	52	39	38	42	44
0	43	42	41	45	46	47	49
-1	50	51	52	44	49	48	47
-2	46	45	44	43	42	41	47
-3	49	48	47	46	44	43	42

	X	Y	A	B	C	D	E
11.	-2	-2	40	45	41	46	44
12.	+1	+2	50	49	48	42	51
13.	-3	-3	49	48	47	42	44
14.	-1	-3	44	47	45	42	41
15.	0	-3	49	47	45	44	46

11. (-2, -2)
    a. 40
    b. 45
    c. 41
    d. 46
    e. 44

12. (1, 2)
    a. 50
    b. 49
    c. 48
    d. 42
    e. 51

13. (-3, -3)
    a. 49
    b. 48
    c. 47
    d. 42
    e. 44

14. (-1, -3)
    a. 44
    b. 47
    c. 45
    d. 42
    e. 41

15. (0, -3)
    a. 49
    b. 47
    c. 45
    d. 44
    e. 46

The next five questions are based on the following table:

	-3	-2	-1	0	+1	+2	+3
+3	56	56	58	55	44	43	42
+2	44	45	55	57	59	58	55
+1	43	42	41	42	40	50	51
0	52	53	57	56	58	59	54
-1	55	51	49	48	45	46	44
-2	42	43	41	39	59	60	61
-3	35	37	38	40	39	33	34

	X	Y	A	B	C	D	E
16.	0	+3	43	45	41	52	55
17.	-1	-3	44	42	51	39	38
18.	-2	-2	55	43	58	45	50
19.	-3	-2	55	43	42	58	56
20.	+1	+3	45	46	55	44	42

16. (0, 3)
    a. 43
    b. 45
    c. 41
    d. 52
    e. 55

17. (-1, -3)
    a. 44
    b. 42
    c. 51
    d. 39
    e. 38

18. (-2, -2)
    a. 55
    b. 43
    c. 58
    d. 45
    e. 50

19. (-3, -2)
    a. 55
    b. 43
    c. 42
    d. 58
    e. 56

20. (1, 3)
    a. 45
    b. 46
    c. 55
    d. 44
    e. 42

The next five questions are based on the following table:

	-3	-2	-1	0	+1	+2	+3
+3	15	12	11	9	2	5	6
+2	7	10	13	14	8	7	6
+1	5	6	3	0	16	12	21
0	16	18	19	4	7	12	10
-1	9	8	12	13	16	17	21
-2	12	11	9	16	19	20	14
-3	7	8	16	17	12	11	9

	X	Y	A	B	C	D	E
21.	+2	+2	11	7	8	9	10
22.	0	+3	15	16	11	9	12
23.	-2	-3	6	7	10	5	8
24.	+1	+1	15	13	16	11	12
25.	+2	+3	10	12	9	5	8

21. (2, 2)
    a. 11
    b. 7
    c. 8
    d. 9
    e. 10

22. (0, 3)
    a. 15
    b. 16
    c. 11
    d. 9
    e. 12

23. (-2, -3)
    a. 6
    b. 7
    c. 10
    d. 5
    e. 8

24. (1, 1)
    a. 15
    b. 13
    c. 16
    d. 11
    e. 12

25. (2, 3)
    a. 10
    b. 12
    c. 9
    d. 5
    e. 8

The next five questions are based on the following table:

	-3	-2	-1	0	+1	+2	+3
+3	77	71	72	76	76	70	65
+2	67	69	68	62	63	61	66
+1	73	74	78	79	64	63	61
0	75	76	72	77	76	79	80
-1	68	71	74	76	72	81	82
-2	59	70	76	73	72	79	78
-3	76	74	72	78	71	70	69

	X	Y	A	B	C	D	E
26.	-1	+3	71	72	73	75	76
27.	-1	0	72	74	76	77	79
28.	-3	0	72	73	76	75	74
29.	0	+2	76	75	70	62	61
30.	+2	+3	59	70	65	62	71

26. (-1, 3)
    a. 71
    b. 72
    c. 73
    d. 75
    e. 76

27. (-1, 0)
    a. 72
    b. 74
    c. 76
    d. 77
    e. 79

28. (-3, 0)
    a. 72
    b. 73
    c. 76
    d. 75
    e. 74

29. (0, 2)
    a. 76
    b. 75
    c. 70
    d. 62
    e. 61

30. (2, 3)
   a. 59
   b. 70
   c. 65
   d. 62
   e. 71

The next five questions are based on the following table:

	-3	-2	-1	0	+1	+2	+3
+3	65	62	60	59	63	67	64
+2	68	70	69	63	62	60	59
+1	58	71	72	67	69	65	58
0	57	56	61	67	68	70	71
-1	64	63	60	59	55	56	53
-2	70	68	69	64	63	62	72
-3	73	75	74	79	78	77	76

	X	Y	A	B	C	D	E
31.	-3	-2	67	68	74	70	71
32.	-1	+2	65	69	71	70	68
33.	+1	0	59	60	70	72	68
34.	+3	+3	73	70	65	64	67
35.	-3	-3	70	64	65	73	62

31. (-3, -2)
   a. 67
   b. 68
   c. 74
   d. 70
   e. 71

32. (-1, 2)
- a. 65
- b. 69
- c. 71
- d. 70
- e. 68

33. (1, 0)
- a. 59
- b. 60
- c. 70
- d. 72
- e. 68

34. (3, 3)
- a. 73
- b. 70
- c. 65
- d. 64
- e. 67

35. (-3, -3)
- a. 70
- b. 64
- c. 65
- d. 73
- e. 62

The next five questions are based on the following table:

	-3	-2	-1	0	+1	+2	+3
+3	33	32	40	39	35	31	38
+2	30	38	41	36	29	28	27
+1	26	32	34	35	37	38	44
0	43	42	44	46	48	50	49
-1	44	39	38	37	29	23	25
-2	24	23	27	26	29	32	31
-3	32	37	36	39	40	41	42

	X	Y	A	B	C	D	E
36.	-3	+2	40	33	32	30	31
37.	+1	+1	36	37	38	39	40
38.	-3	0	41	43	44	40	39
39.	+2	+3	24	31	26	25	33
40.	-1	-3	36	44	40	43	41

36. (-3, 2)
    a. 40
    b. 33
    c. 32
    d. 30
    e. 31

37. (1, 1)
    a. 36
    b. 37
    c. 38
    d. 39
    e. 40

38. (-3, 0)
    a. 41
    b. 43
    c. 44
    d. 40
    e. 39

39. (2, 3)
    a. 24
    b. 31
    c. 26
    d. 25
    e. 33

40. (-1, -3)
    a. 36
    b. 44
    c. 40
    d. 43
    e. 41

Answer Guide to Table Reading Practice Test

1. **B.** On the x-axis (horizontal), look at the -3 column, then move down until you reach 2 on the y-axis (vertical) column. This number is 28. (-2,3) is 22. (3,1) is 21.(-1,0) is 31. (-1,3) is 23. Keep in mind that some of the numbers on the table are repeated, so different x-values and y-values could be the same number. For instance, (-1, 3) and (-2, -1) are both 23.

2. **A.** The x-value is +3, and the y-value is -2; hence the answer is 23. (-3,2) is 28. (2,-3) is 27. (-2,1) is 29. (0,1) is 25.

3. **C.** The x-value is -2, and the y-value is -1; hence the answer is 23. (-1,1) is 28. (1,2) is 24. (0,1) is 25. (2,1) is 22.

4. **E.** The x-value is +2, and the y-value is +2; hence the answer is 29. (1,3) is 27. (2,-2) is 19. (1,0) is 22. (2,0) is 24.

5. **A.** The x-value is 0, and the y-value is +1; hence the answer is 25. (-1,0) is 31. (0,-1) is 19. (2,1) is 22. (2,-1) is 28.

6. **D.** The x-value is 0, and the y-value is +3; hence the answer is 37. (-2,3) is 32. (-3,0) is 26. (-2,2) is 33. (-1,3) is 35.

7. **B.** The x-value is -2, and the y-value is -1; hence the answer is 33. (-1,2) is 34. (1,2) is 37. (2,1) is 29. (2,-1) is 36.

8. **B.** The x-value is -1, and the y-value is -1; hence the answer is 39. (0,-2) is 37. (1,1) is 28. (0,0) is 29. (-3,3) is 30.

9. **E.** The x-value is +2, and the y-value is +3; hence the answer is 40. (-2,-1) is 33. (3,2) is 31. (2,2) is 32. (-1,2) is 34.

10. **B.** The x-value is 0, and the y-value is +2; hence the answer is 36. (-2,0) is 27. (2,0) is 23. (1,0) is 25. (3,1) is 24.

11. **B.** The x-value is -2, and the y-value is -2; hence the answer is 45. (2,2) is 40. (2,3) is 41. (-2,2) is 46. (0,2) is 44.

12. **D.** The x-value is +1, and the y-value is +2; hence the answer is 42. (-3,-1) is 50. (1,-1) is 49. (2,-1) is 48. (-2,-1) is 51.

13. **A**. The x-value is -3, and the y-value is -3; hence the answer is 49. (2,-1) is 48. (3,-2) is 47. (-2,0) is 42. (-1,-2) is 44.

14. **B**. The x-value is -1, and the y-value is -3; hence the answer is 47. (-3,3) is 44. (-2,-2) is 45. (1,2) is 42. (2,-2) is 41.

15. **E**. The x-value is 0, and the y-value is -3; hence the answer is 46. (3,2) is 49. (2,0) is 47. (0,0) is 45. (0,-1) is 44.

16. **E**. The x-value is 0, and the y-value is +3; hence the answer is 55. (-3,1) is 43. (-2,2) is 45. (-1,-2) is 41. (-3,0) is 52.

17. **E**. The x-value is -1, and the y-value is -3; hence the answer is 38. (-3,2) is 44. (-2,1) is 42. (3,1) is 51. (1,-3) is 39.

18. **B**. The x-value is -2, and the y-value is -2; hence the answer is 43. (-1,2) is 55. (2,2) is 58. (-2,2) is 45. (2,1) is 50.

19. **C**. The x-value is -3, and the y-value is -2; hence the answer is 42. (-1,2) is 55. (2,3) is 43. (1,0) is 58. (-2,3) is 56.

20. **D**. The x-value is +1, and the y-value is +3; hence the answer is 44. (-2,2) is 45. (2,-1) is 46. (-3,-1) is 55. (-2,1) is 42.

21. **B**. The x-value is +2, and the y-value is +2; hence the answer is 7. (-2,-2) is 11. (-2,-1) is 8. (-3,-1) is 9. (3,0) is 10.

22. **D**. The x-value is 0, and the y-value is +3; hence the answer is 9. (-3,3) is 15. (-3,0) is 16. (-1,3) is 11. (-1,-1) is 12.

23. **E**. The x-value is -2, and the y-value is -3; hence the answer is 8. (3,2) is 6. (2,2) is 7. (-2,2) is 10. (2,3) is 5.

24. **C**. The x-value is +1, and the y-value is +1; hence the answer is 16. (-3,3) is 15. (-1,2) is 13. (2,-3) is 11. (-1,-1) is 12.

25. **D**. The x-value is +2, and the y-value is +3; hence the answer is 5. (-2,2) is 10. (-3,-2) is 12. (3,-3) is 9. (1,2) is 8.

26. **B.** The x-value is -1, and the y-value is +3; hence the answer is 72. (-2,3) is 71. (-3,1) is 73. (-3,0) is 75. (1,0) is 76.

27. **A.** The x-value is -1, and the y-value is 0; hence the answer is 72. (-2,1) is 74. (0,-1) is 76. (0,0) is 77. (0,1) is 79.

28. **D.** The x-value is -3, and the y-value is 0; hence the answer is 75. (-1,0) is 72. (-3,1) is 73. (0,3) is 76. (-2,1) is 74.

29. **D.** The x-value is 0, and the y-value is +2; hence the answer is 62. (-2,0) is 76. (-3,0) is 75. (2,-3) is 70. (3,1) is 61.

30. **B.** The x-value is +2, and the y-value is +3; hence the answer is 70. (-3,-2) is 59. (3,3) is 65. (0,2) is 62. (1,-3) is 71.

31. **D.** The x-value is -3, and the y-value is -2; hence the answer is 70. (2,3) is 67. (-2,-2) is 68. (-1,-3) is 74. (3,0) is 71.

32. **B.** The x-value is -1, and the y-value is +2; hence the answer is 69. (2,1) is 65. (-2,1) is 71. (2,0) is 70. (-3,2) is 68.

33. **E.** The x-value is +1, and the y-value is 0; hence the answer is 68. (0,-1) is 59. (-1,-1) is 60. (2,0) is 70. (3,-2) is 72.

34. **D.** The x-value is +3, and the y-value is +3; hence the answer is 64. (-3,-3) is 73. (-3,-2) is 70. (2,1) is 65. (0,0) is 67.

35. **D.** The x-value is -3, and the y-value is -3; hence the answer is 73. (2,0) is 70. (3,3) is 64. (-3,3) is 65. (1,2) is 62.

36. **D.** The x-value is -3, and the y-value is +2; hence the answer is 30. (1,-3) is 40. (-3,3) is 33. (-2,3) is 32. (2,3) is 31.

37. **B.** The x-value is +1, and the y-value is +1; hence the answer is 37. (0,2) is 36. (-1,-1) is 38. (-2,-1) is 39. (1,-3) is 40.

38. **B.** The x-value is -3, and the y-value is 0; hence the answer is 43. (-1,2) is 41. (-1,0) is 44. (1,-3) is 40. (0,3) is 39.

39. **B.** The x-value is +2, and the y-value is +3; hence the answer is 31. (-3,-2) is 24. (0,-2) is 26. (3,-1) is 25. (-3,3) is 33.

40. **A**. The x-value is -1, and the y-value is -3; hence the answer is 36. (3,1) is 44. (1,-3) is 40. (-3,0) is 43. (2,-3) is 41.

INSTRUMENT COMPREHENSION

The AFOQT will test your ability to comprehend important instruments; it will contain questions regarding the direction and orientation of a plane based on artificial horizon dials and compasses. The pictures of these instruments can look slightly different from test to test, but they will be read in the same way that you will learn about in this section.

How to Read the Artificial Horizon Dial and Compass

The artificial horizon dial, also referred to as the *attitude indicator,* is an instrument that exhibits the plane's orientation in relation to the ground. It shows if the nose of the aircraft is pitching (tilting) up or pitching down in relation to the Earth's horizon. It also shows if the wings are banking (tilting) left or banking right. These dials are read in the cockpit of the plane, so it can help to visualize these dials as if you were the pilot of the aircraft. In the pictures on the next page, you will notice how the artificial horizon dial has a small silhouette of a plane or wings in the center. The dark line across the dial separates a lighter shade and a darker shade. The darker shade is the earth, and the lighter shade is the sky. The dark line that divides the earth from the sky is the horizon. Keep in mind that other artificial horizon dials may look different, but the important thing is to pay attention to the horizon bar that separates the sky from the ground. If the silhouette is above the horizon bar, then the plane is pitching up and heading farther into the sky **(Figure A)**. If the silhouette is under the horizon bar, then the plane is pitching down and headed toward the earth **(Figure B)**. If the silhouette is directly on the horizon bar, then the aircraft is in level flight, neither pitching up nor down **(Figure C)**. Besides being above, below, or even with the horizon bar, the silhouette shows if the plane is banking left or right. If the silhouette's left wing is tilted toward the earth, the plane is banking left **(Figure D)**. If the silhouette's right wing is tilted toward the earth, the plane is banking right **(Figure E)**. If the silhouette is neither banking left or right, then the silhouette's wings will be even **(Figure F)**. Keep in mind that a plane could be pitching up or down while banking left or right. To accurately read the dials, it is important to ask yourself two things: "Is the plane silhouette above or below the horizon bar?" and "Is the plane silhouette banking left, right, or neither?"

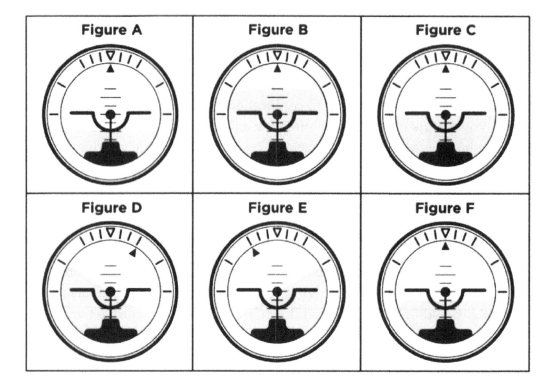

The compass may be a familiar tool to you. When the needle is pointing straight up toward the "N," the aircraft is headed north. If the needle is pointed down at the "S," the aircraft is headed south. If the needle is between the "N" and the "W," the aircraft is headed northwest. When trying to identify the orientation of the plane with each given compass, think of north as in front of you, south as behind you, west as to your left, and east as to your right. This is how the images on the test will be oriented. If a plane were in front of you and heading south in level flight, it would be headed right at you. If it were heading north in level flight, you would see its rear as it flew away from you. If it were heading west, it would be flying away on your left side. If it were heading east, it would be flying away on your right side. If a plane in front of you were heading "SW" (southwest), the plane would appear as if it were heading to your left, yet slightly toward you.

At first, these dials may seem complicated, but after some practice, it should be simple to identify the orientation of the plane with each set of dials. Consider the following example. You will notice an artificial horizon dial to the far left and a compass to its right. Determine if the nose of the aircraft is pitched up or down (is the silhouette above or below the horizon bar?). Then determine if the plane is banking to the left or right. Next, look at the compass and identify the aircraft's direction. North is away from you and south is toward you. West is to your left and east is to your right. When determining the plane's orientation, it may help to imagine yourself outside the airplane, viewing its direction.

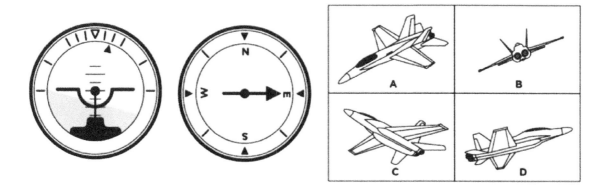

The correct answer is **D**. First, examine the attitude indicator. The plane silhouette is above the horizon bar and a couple notches into the white. This indicates that the aircraft is ascending slightly. The silhouette is also slightly tilted to the left, so we know it is banking slightly left. Now examine the compass. The arrow is pointed east. This means the aircraft will be flying toward our right. The ascent, eastbound heading, and banking left are all exemplified by option **D**.

If you want more practice with reading instruments, working with a compass can be helpful. You can buy an affordable one online, or get a smartphone app. An attitude indicator would be difficult to locate so doing the following practice test is a good way to become proficient.

Before You Start the Practice Test

On the next page, you will begin an Instrument Comprehension practice test. Set a timer for 5 minutes before you start this practice test. Giving yourself 5 minutes will give you an authentic feel for how long you have to finish this portion of the AFOQT. Like the official test, this practice test has exactly 25 items. Set your timer, turn the page, and begin.

Instrument Comprehension Practice Test Questions

(5 minutes)

1. Based on the dials below, which selection accurately displays the orientation of the plane?

1

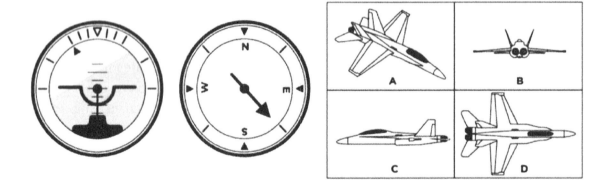

2. Based on the dials below, which selection accurately displays the orientation of the plane?

2

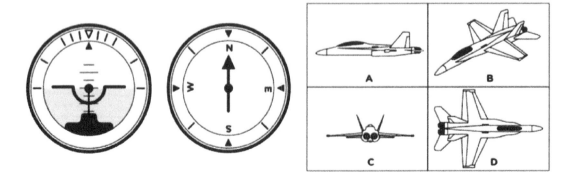

3. Based on the dials below, which selection accurately displays the orientation of the plane?

3

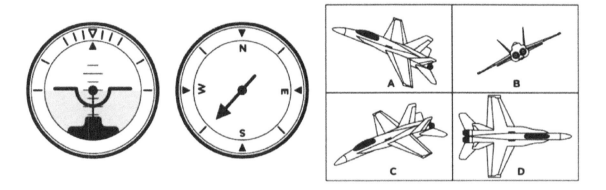

4. Based on the dials below, which selection accurately displays the orientation of the plane?

4

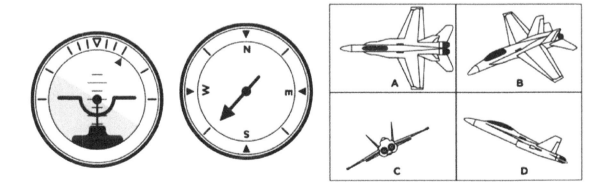

5. Based on the dials below, which selection accurately displays the orientation of the plane?

5

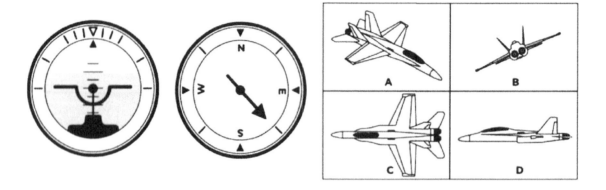

6. Based on the dials below, which selection accurately displays the orientation of the plane?

6

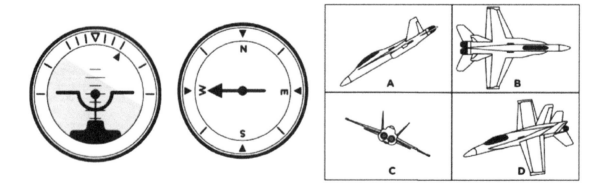

7. Based on the dials below, which selection accurately displays the orientation of the plane?

7

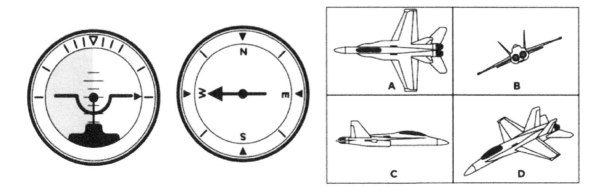

8. Based on the dials below, which selection accurately displays the orientation of the plane?

8

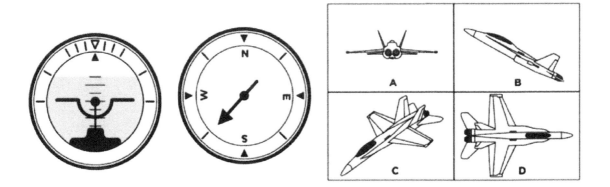

9. Based on the dials below, which selection accurately displays the orientation of the plane?

9

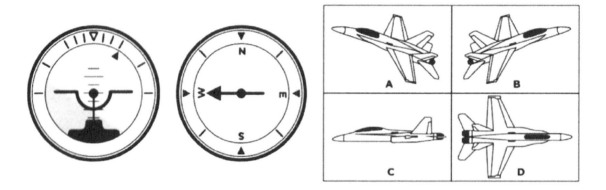

10. Based on the dials below, which selection accurately displays the orientation of the plane?

10

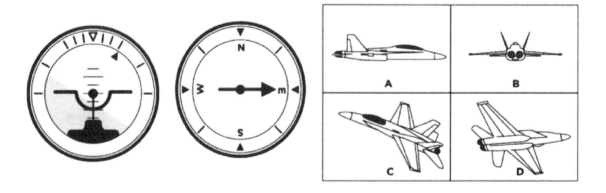

11. Based on the dials below, which selection accurately displays the orientation of the plane?

11

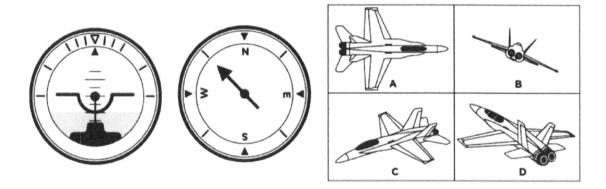

12. Based on the dials below, which selection accurately displays the orientation of the plane?

12

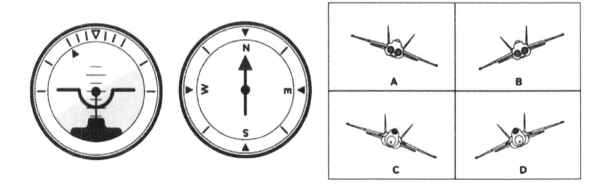

13. Based on the dials below, which selection accurately displays the orientation of the plane?

13

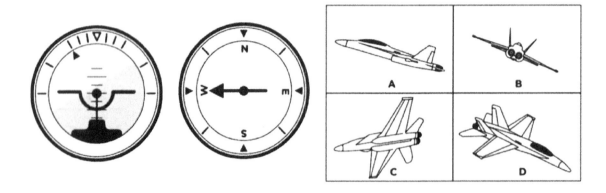

14. Based on the dials below, which selection accurately displays the orientation of the plane?

14

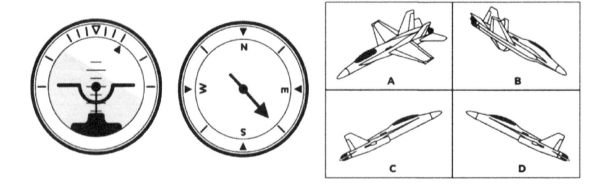

15. Based on the dials below, which selection accurately displays the orientation of the plane?

15

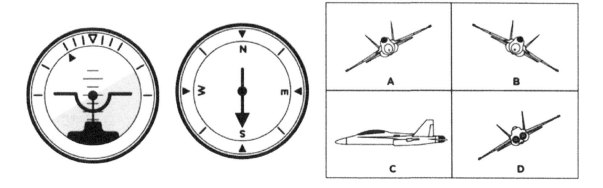

16. Based on the dials below, which selection accurately displays the orientation of the plane?

16

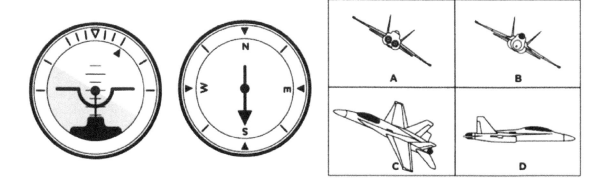

17. Based on the dials below, which selection accurately displays the orientation of the plane?

17

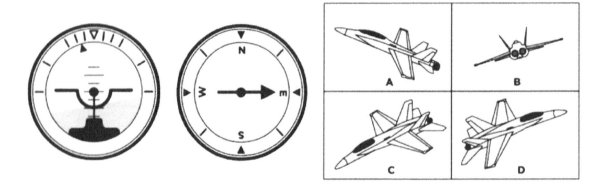

18. Based on the dials below, which selection accurately displays the orientation of the plane?

18

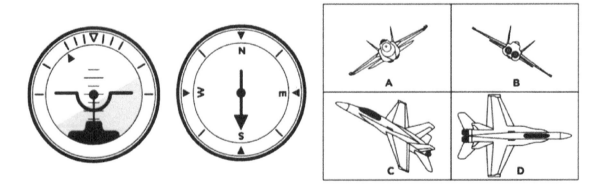

19. Based on the dials below, which selection accurately displays the orientation of the plane?

19

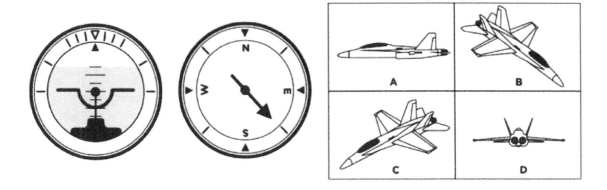

20. Based on the dials below, which selection accurately displays the orientation of the plane?

20

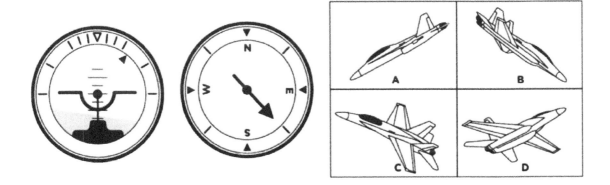

21. Based on the dials below, which selection accurately displays the orientation of the plane?

21

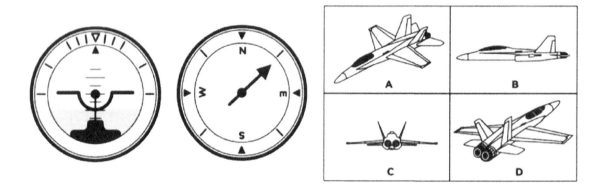

22. Based on the dials below, which selection accurately displays the orientation of the plane?

22

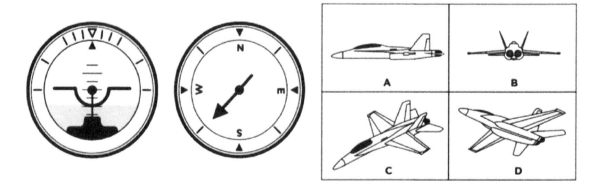

23. Based on the dials below, which selection accurately displays the orientation of the plane?

23

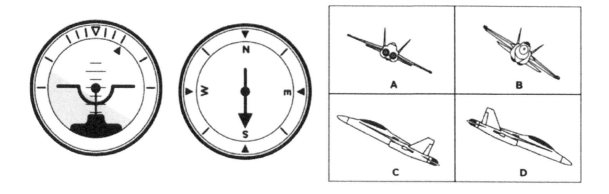

24. Based on the dials below, which selection accurately displays the orientation of the plane?

24

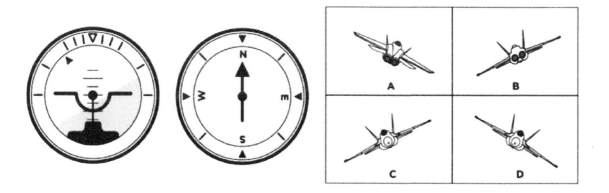

25. Based on the dials below, which selection accurately displays the orientation of the plane?

25

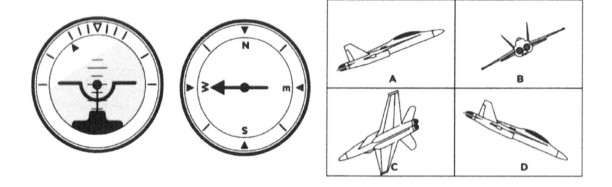

Answer Guide to Instrument Comprehension Practice Test

1. **A.** The airplane is heading southeast, descending, and banking right.
2. **C.** The airplane is heading north, in level flight, and not banking.
3. **C.** The airplane is heading southwest, in level flight, and not banking.
4. **B.** The airplane is heading southwest, in level flight, and banking left.
5. **A.** The airplane is heading southeast, descending, and not banking.
6. **D.** The airplane is heading west, descending, and banking left.
7. **A.** The airplane is heading west, in level flight, and banking left.
8. **C.** The airplane is heading southwest, descending, and not banking.
9. **A.** The airplane is heading west, ascending, and banking left.
10. **D.** The airplane is heading east, ascending, and banking left.
11. **D.** The airplane is heading northwest, ascending, and not banking.
12. **A.** The airplane is heading north, in level flight, and banking right,
13. **C.** The airplane is heading west, descending, and banking right.
14. **B.** The airplane is heading southeast, descending, and banking left.
15. **A.** The airplane is heading south, in level flight, and banking right.
16. **B.** The airplane is heading south, in level flight, and banking left.
17. **D.** The airplane is heading east, ascending, and banking right.
18. **A.** The airplane is heading south, ascending, and banking right.
19. **B.** The airplane is heading southeast, descending, and not banking.
20. **D.** The airplane is heading southeast, ascending, and banking left.
21. **D.** The airplane is heading northeast, ascending, and not banking.
22. **D.** The airplane is heading southwest, ascending, and not banking.
23. **B.** The airplane is heading south, ascending, and banking left.
24. **A.** The airplane is heading north, ascending, and banking right.
25. **C.** The airplane is heading west, descending, and banking right.

BLOCK COUNTING

The Block Counting portion of the AFOQT tests your logical, geometric, and spatial reasoning. It is not necessary to have prior experience with block counting, but this section will prepare you so that you know what to expect. During the test, you will be shown three-dimensional block drawings. Each question will ask how many blocks are touching one specific block. By analyzing each figure of blocks, you will determine how many are touching that block. Some of the blocks may not be visible, but you will use your reasoning to determine how many blocks are touching. The blocks will typically be configured in irregular formations.

Determining the Number of Blocks That Touch

As you observe the block formation below, it is important to note that only faces count toward the total number of blocks that touch. Corners are not considered to be touching in this test. For instance, block A is touching four blocks (block numbers 5, 4, 6 and 2). Blocks 1, 3, 8, and 7 are not touching block A. Corners do not count.

Block 1 is touching two blocks (4 and 5). Block 4 is touching three blocks (1, A, and 3). Block 3 is touching two blocks (4 and 6). Block 5 is touching three blocks (1, A, and 8). Block 6 is touching three blocks (3, A, and 7). Block 8 is touching two blocks (5 and 2). Block 2 is touching three blocks (8, A, and 7). Block 7 is touching two blocks (2 and 6).

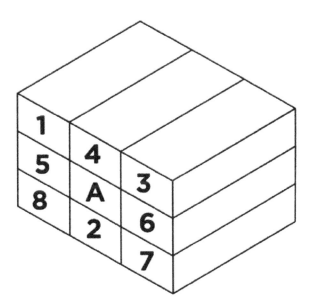

In the figure below, determine how many blocks are touching block A:

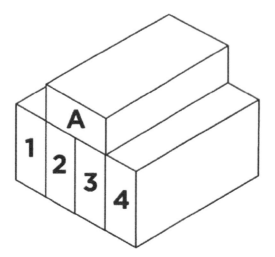

Answer: Two blocks are touching block A (blocks 2 and 3). Blocks 1 and 4 are not touching block A (edges/corners do not count). Even though you may consider blocks 1 and 4 to be touching block A, they do not touch in respect to the parameters of the test. Only faces are counted as touching, not edges.

Tips for Counting Blocks

If you are having trouble visualizing the number of blocks that are touching the specified block, it may help to think about each face of the block and how many other blocks are touching each individual face. Try to use this method with the block below. How many blocks are touching block A? Think of each face of block A and count how many blocks are touching each face:

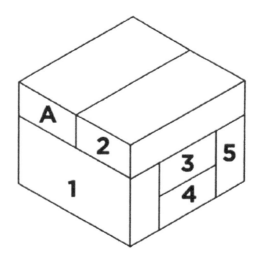

Answer: In total, four blocks are touching block A. One block is touching the right face of block A. Three blocks are touching the bottom face of block A. No blocks are touching the front, back, left, or top faces of block A.

Understanding the Directional Orientation of Each Figure

In the answer key, terms such as "right," "left," "top/above/over," "under/bottom/below," "behind/back," and "front" are used. To clear up potential confusion, we created the diagram below to define each direction in reference to this section:

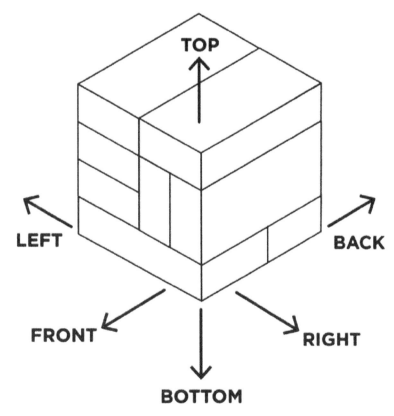

Before You Start the Practice Test

On the next page, you will begin a Block Counting practice test. Set a timer for 4.5 minutes before you start this practice test. Giving yourself 4.5 minutes will give you an authentic feel for how long you have to finish this portion of the AFOQT. Like the official test, this practice test has exactly 30 items. Set your timer, turn the page, and begin.

Block Counting Practice Test Questions
(4.5 Minutes)

Instructions: For each question (1 through 30), determine how many blocks the specified block is touching. Assume all blocks are the same size.

Use the figure below for questions 1-5:

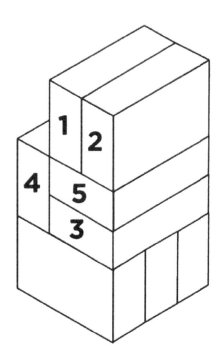

1. Block 1
    a. 1
    b. 2
    c. 3
    d. 4
    e. 5

2. Block 2
    a. 1
    b. 2
    c. 3
    d. 4
    e. 5

3. Block 3
    a. 4
    b. 5
    c. 6
    d. 7
    e. 8

4. Block 4
    a. 2
    b. 3
    c. 4
    d. 5
    e. 6

5. Block 5
    a. 2
    b. 3
    c. 4
    d. 5
    e. 6

Use the figure below for questions 6-10:

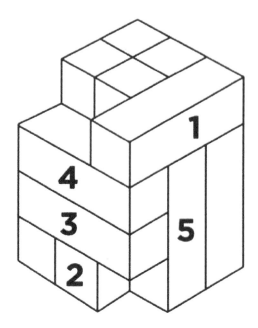

6. Block 1
    a. 2
    b. 3
    c. 4
    d. 5
    e. 6

7. Block 2
    a. 4
    b. 5
    c. 6
    d. 7
    e. 8

8. Block 3
    a. 6
    b. 7
    c. 3
    d. 4
    e. 5

9. Block 4
    a. 7
    b. 5
    c. 6
    d. 8
    e. 4

10. Block 5
    a. 6
    b. 7
    c. 8
    d. 5
    e. 9

Use the figure below for questions 11-15:

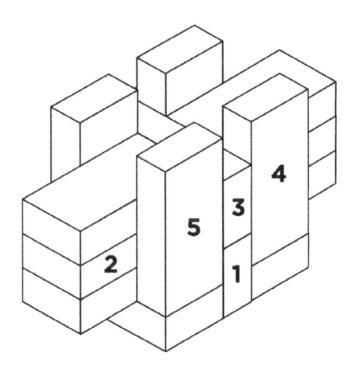

11. Block 1
    a. 9
    b. 7
    c. 5
    d. 8
    e. 4

12. Block 2
    a. 4
    b. 5
    c. 8
    d. 9
    e. 5

13. Block 3
    a. 7
    b. 9
    c. 6
    d. 4
    e. 8

14. Block 4
    a. 3
    b. 5
    c. 6
    d. 7
    e. 9

15. Block 5
    a. 7
    b. 8
    c. 9
    d. 5
    e. 6

Use the figure below for questions 16-20:

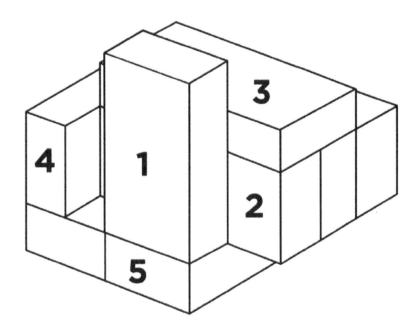

16. Block 1
    a. 6
    b. 7
    c. 3
    d. 4
    e. 5

17. Block 2
    a. 4
    b. 5
    c. 6
    d. 8
    e. 7

18. Block 3
    a. 1
    b. 2
    c. 3
    d. 4
    e. 5

19. Block 4
    a. 6
    b. 7
    c. 3
    d. 2
    e. 4

20. Block 5
    a. 7
    b. 6
    c. 5
    d. 4
    e. 3

Use the figure below for questions 21-25:

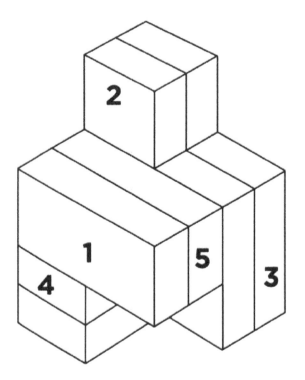

21. Block 1
    a. 1
    b. 2
    c. 3
    d. 4
    e. 5

22. Block 2
    a. 2
    b. 3
    c. 4
    d. 6
    e. 5

23. Block 3
    a. 4
    b. 3
    c. 5
    d. 6
    e. 2

24. Block 4
    a. 5
    b. 7
    c. 4
    d. 3
    e. 8

25. Block 5
    a. 3
    b. 6
    c. 5
    d. 7
    e. 4

Use the figure below for questions 26-30:

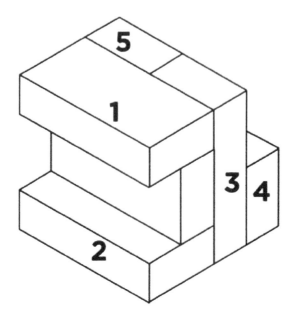

26. Block 1
    a. 1
    b. 2
    c. 3
    d. 4
    e. 5

27. Block 2
    a. 1
    b. 2
    c. 3
    d. 4
    e. 5

28. Block 3
    a. 1
    b. 2
    c. 3
    d. 4
    e. 5

29. Block 4
    a. 1
    b. 2
    c. 3
    d. 4
    e. 5

30. Block 5
    a. 1
    b. 2
    c. 3
    d. 4
    e. 5

Answer Guide to Block Counting Practice Test

1. **B.** Block 1 touches two blocks: block 2 and block 5.
2. **B.** Block 2 touches two blocks: block 1 and block 5.
3. **B.** Block 3 touches five blocks: block 5, block 4, and three blocks underneath
4. **D.** Block 4 touches five blocks: block 5, block 3, and three blocks underneath.
5. **C.** Block 5 touches four blocks: block 1, block 2, block 4, and block 3.
6. **D.** Block 1 touches five blocks: block 4, block 5, one block behind block 5, two blocks to the left of block 1.
7. **C.** Block 2 touches six blocks: block 3, block 5, one block behind block 5, two blocks behind block 3, and one block to the left of block 2.
8. **A.** Block 3 touches six blocks: block 4, block 2, block 5, two blocks behind block 3, and one block to the left of block 2.
9. **B.** Block 4 touches five blocks: block 1, block 5, block, 3, and two blocks behind block 4.
10. **A.** Block 5 touches six blocks: block 1, block 4, block 3, block 2, one block behind block 5, and one block to the left of block 5.
11. **A.** Block 1 touches nine blocks: block 3, block 5, block 4, three unmarked blocks in front, and three unmarked blocks behind.
12. **B.** Block 2 touches five blocks: block 5, block 3, one block under and one block over, and one block to the left.
13. **B.** Block 3 touches nine blocks: block 1, block 2, block 5, block 4, two unmarked blocks in front, and three unmarked blocks behind.
14. **C.** Block 4 touches six blocks: block 3, block 1, three unmarked blocks to the left, and one on the bottom.
15. **E.** Block 5 touches six blocks: block 1, block 3, block 2, two unmarked blocks under and over block 2, and one unmarked block on the bottom.
16. **C.** Block 1 touches three blocks: block 3, block 2, and block 5.
17. **C.** Block 2 touches six blocks: block 3, block 4, block 1, block 5, one unmarked block on the bottom and one block behind.
18. **C.** Block 3 touches three blocks: block 1, block 2, and one unmarked block underneath.
19. **E.** Block 4 touches four blocks: block 2, two unmarked blocks behind block 2, and one unmarked block on the bottom.
20. **C.** Block 5 touches five blocks: block 1, block 2, one unmarked block to the left, and two unmarked blocks behind block 2.
21. **B.** Block 1 touches two blocks: block 4 and block 5.
22. **C.** Block 2 touches four blocks: block 5, block 4, one unmarked block to the right, and one unmarked block behind.
23. **A.** Block 3 touches four blocks: block 4, one unmarked block under block 4, one unmarked block in front of block 3, and one unmarked block to the left.

24. B. Block 4 touches seven blocks: block 1, block 5, block 2, block 3, one unmarked block on top, one unmarked block on bottom, and one unmarked block to the right.
25. E. Block 5 touches four blocks: block 1, block 4, block 2, and one unmarked block behind.
26. C. Block 1 touches three blocks: block 5, block 3, and one unmarked block below.
27. C. Block 2 touches three blocks: block 3, block 5, and one unmarked block above.
28. E. Block 3 touches five blocks: block 4, block 1, block 5, block 2, and one unmarked block in front.
29. B. Block 4 touches two blocks: block 3 and block 5.
30. E. Block 5 touches five blocks: block 1, block 2, block 3, block 4, and the unmarked block in front.

AVIATION INFORMATION

This section of the AFOQT will test your knowledge of aviation. If you are interested in becoming a pilot or getting a position such as Air Battle Manager, it is recommended that you seek out additional comprehensive study materials that exclusively focus on aviation. In this study guide, basic information of the following topics will be covered:

- Aerodynamics and flight terms
- Fixed-wing aircraft
- Flight controls
- Four basic flight maneuvers
- Flight envelope
- Important airport information
- Helicopters

Aerodynamics and Flight Terms

The study of moving air and solid bodies moving through it is called aerodynamics. This is most relevant in the field of aviation since an aircraft moves through the air and several forces affect motion.

As we navigate these different forces, it is important to note the role of atmospheric pressure. Atmospheric pressure is affected by the altitude above sea level, and air density is inversely proportional to atmospheric pressure. So as the aircraft climbs higher, the density of the air decreases. As the aircraft descends, the air density increases. Also, the amount of humidity in the air changes with temperature.

Now let's explore the four forces of flight: gravity, drag, lift, and thrust. For the aircraft to fly, the lift and thrust of the aircraft must be greater than the drag and gravitational pull.

Gravity is the force that moves the aircraft downward. It attracts the aircraft to the center of the earth at a rate of 9.8 m/s^2. Gravity acts on the mass of the aircraft, resulting in weight. Lift must be greater than weight for flight to occur.

Lift is the force that moves the aircraft upward. It opposes weight. As the aircraft goes faster, more lift occurs. The air that goes above the curved surface of the wing has farther to go than the air that goes underneath the flat surface of the wing. Since the air above the wing must travel faster, the air pressure above the wing is lower. Below the wing, where the air is traveling more slowly, the air pressure is higher. This difference in air pressure above and below the wing creates lift. This occurrence is explained by the *Bernoulli principle*.

Drag is the force that moves the aircraft backward. When an aircraft is moving through the air, aerodynamic friction or wind resistance occurs. The shape of the aircraft, the speed at which it is traveling, and the density of the air that it is passing through all contribute to the amount of drag. Two common types of drag exist in flight: *induced drag* and *parasitic drag*. Induced drag involves the friction caused by lift. Induced drag increases as lift increases. Parasitic drag is the result of an aircraft moving through the air. As the speed of the aircraft increases, the parasitic drag increases. This requires more thrust from the engine to overcome drag.

Thrust is the force that moves the aircraft forward. It opposes drag. The jet engines or propellers of the aircraft create thrust and propel the aircraft forward. The thrust must overcome the drag and in turn provides airflow beneath the wings for lift to occur.

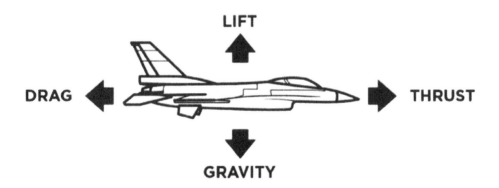

Flight attitude is the inclination of the three principal axes of an aircraft: longitudinal, lateral, and vertical axes.

The **longitudinal axis** extends through the rear tail and front nose of the aircraft. Movement around the longitudinal axis is known as **roll** and is controlled by the ailerons.

The **lateral axis** extends out through the left and right of the aircraft, perpendicular to the aircraft. Movement along this axis is known as **pitch** and is controlled by the elevators.

The **vertical axis** extends out the top and bottom of the aircraft, perpendicular to the other two axes. Movement along the vertical axes is known as **yaw** and is controlled by the rudder.

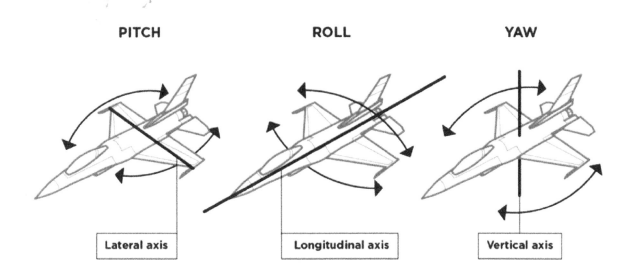

Fixed-Wing Aircraft

Fixed-wing aircraft are distinct from rotary-wing aircraft (helicopters) in that forward motion and the shape of their wings generate lift, rather than the movement of the wings themselves. The wings may move a little during flight, but the movement harnesses the power of airflow to generate lift rather than generating lift on its own, such as a helicopter does.

Fixed-wing wing aircraft have the following six basic parts: wings, fuselage, tail assembly, landing gear, powerplant, and flight controls.

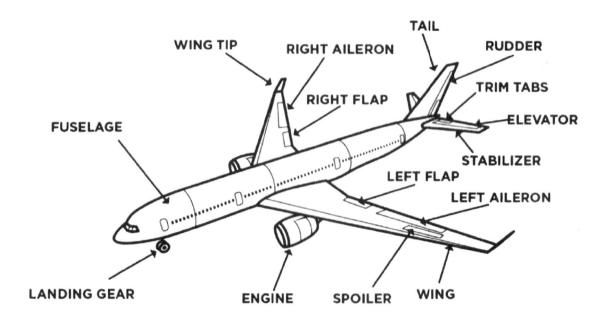

Wings

The wings are the main *airfoils* of the plane. Airfoils are surfaces designed to generate lift when moving through the air. They are shaped with a rounded leading edge and a sharp trailing edge. With a rounded upper surface and a relatively flat undersurface, the airflow moves faster above the wings and slower beneath the wings. This difference in speed of airflow generates greater air pressure below the wings, resulting in lift.

The flaps and ailerons are attached to the wings. Typically, the flaps are on the inner trailing edge of the wings and the ailerons are on the outer trailing edge of the wings. Flaps can help change the *camber* of the wings to generate more lift. Camber is a term used to describe the curvature of the airfoils. If wings, or airfoils, have high camber, then they have a greater curvature than wings with low camber. The *mean camber line* is an imaginary line that equally divides the upper and lower parts of the wing, so that the thickness above and below the mean camber line is equal. This is not to be confused with the *chord line,* which is an imaginary line that runs straight from the leading edge of the airfoil to the trailing edge of the airfoil.

Wings are commonly referred to as cantilever wings or semi-cantilever wings. Cantilever wings are attached directly to the fuselage with no need for external bracing. The internal structure of the fuselage is strong enough to support cantilever wings, but semi-cantilever wings require external bracing for support.

Fuselage

The fuselage is the body of the aircraft that includes the cabin, cockpit, and cargo area. It has attachment points for the empennage (tail assembly), engines, and the wings. Fuselage structures can be referred to as *truss, monocoque,* and *semi-monocoque.* A truss fuselage structure is typically made of metal tubing welded together to form the skeleton of the aircraft. A monocoque fuselage is a single shell with its primary structure being its outer surface. A semi-monocoque fuselage is essentially a blend of truss and monocoque fuselage structures. Semi-monocoque is very common today and includes both a metal frame and outer surface.

Tail Assembly

The tail assembly, also known as the empennage, is a structure at the rear of the aircraft that stabilizes the aircraft during flight. It incorporates fixed vertical and horizontal stabilizers as well as the rudder and elevators to control pitch and yaw. The rudder is attached to the vertical stabilizer, and the elevators are attached to the horizontal stabilizers. Some aircraft have trim tabs on the ear of the rudder and elevators to assist with small adjustments during flight. The rear end of the fuselage that the vertical and horizontal stabilizers are connected to is considered part of the tail assembly.

Landing Gear

The landing gear assists the airplane while on the ground. It typically consists of wheels, but it can consist of floats if landing on water, or skis if landing in snow. The wheels not only assist with landing but also with moving on the ground and helping with takeoff. Often, landing gear is retracted into the body of the aircraft to reduce drag during flight. Sometimes, with smaller airplanes, landing gear is not retractable.

Two common types of landing gear configurations are known as *conventional gear* and *tricycle gear*. Conventional gear, also known as *traildagger*, consists of two main wheels toward the front of the aircraft and one much smaller wheel toward the rear of the aircraft. *Tricycle gear* is the most common configuration and consists of two sets of wheels side by side underneath the wings and a smaller wheel toward the nose of the aircraft.

Powerplant

An aircraft powerplant is essentially its engine (or engines). It is a component of the propulsion system necessary to produce thrust to propel the aircraft forward.

Flight Controls

The **vertical stabilizer**, or fin, helps stabilize the aircraft against horizontal wind. The vertical stabilizer pushes against the side with the most air pressure to prevent yaw and stabilize the aircraft.

The **horizontal stabilizer** helps stabilize the plane by preventing up and down movement, or pitching, of the aircraft nose.

Flight control systems are subdivided into *primary* and *secondary controls.* The primary control surfaces are the ailerons, rudder, and elevators. The secondary flight control systems include, but are not limited to, the flaps, slats, slots, spoilers, speed brakes, and trim control systems.

Primary

Earlier, we discussed the movements of pitch, yaw, and roll. These movements are handled by the aircraft's primary controls.

The **ailerons** control the rotation around the longitudinal axis of an aircraft. This is the roll or bank motion of an aircraft. They are on the outboard trailing edge of the wings. The ailerons move in opposite directions of one another. For example, if the ailerons on the right wing went down, and the ailerons on the left wing went up, the aircraft would start to roll left.

The **elevators** control the rotation around the lateral axis of the aircraft. This is the pitch motion of the aircraft. The elevators are attached to the horizontal stabilizer.

The **rudder** controls the rotation around the vertical axis. This is the yaw motion of the aircraft. The rudder is attached to the vertical stabilizer, or tail fin. It moves side to side, pushing the tail left or right. The rudder is used in conjunction with the ailerons to help prevent yaw.

Secondary

Flaps are the most common flight control. They are on the trailing edge of the wings, typically between the fuselage and the ailerons. When deployed, they can increase the camber of the wing, thus increasing its curvature. During takeoff, flaps can help create more lift. They are often retracted mid-flight since they can increase drag.

Speed brakes are most common on military jets. They help reduce speed during landing.

Spoilers are plates on top of the wings. They can be raised to reduce lift and increase drag. They are often used during landings to decrease skidding, which helps with ground effect. Spoilers can help keep the aircraft down on its wheels and reduce drag. Speed brakes do not affect lift, so spoilers are deployed in order to reduce it. Spoilers can also be used to control roll during flight and increase the descent rate.

Slats are surfaces on the leading edge of wings. When deployed, there exists a gap between the wing and the slat, allowing air to flow from under the wing to over the wing surface. Slats allow enough lift while flying at higher angles of attack, making it possible to land in a shorter distance and fly at lower speeds without stalling.

Slots are span-wise gaps in each wing that allow air to flow from under the wing to over the wing. This provides lift at a higher angle of attack than what would be possible without slots.

Trim systems help stabilize the aircraft in a desired attitude position without the pilot having to constantly apply control force. They are small surfaces connected to the trailing edge of primary flight controls such as ailerons, elevators, or rudders.

Related terms:

Angle of attack describes the angle between the relative oncoming wind and chord line (the line between the leading and trailing edges of the airfoil).

Ground effect is the increased lift and decreased drag that occurs when the aircraft's wings are in close proximity to a fixed surface or the ground.

Four Basic Flight Maneuvers

The four basic flight maneuvers are straight-and-level, turn, climb, and descent. Every flight typically contains a combination of these four fundamental maneuvers.

Straight-and-Level Flight
As the name implies, straight-and-level flight maintains a constant heading and altitude. Any deviation in heading or altitude from unintentional climbs, descents, or turns is immediately corrected. Different methods are used to maintain straight-and-level flight. One method is to use the attitude indicator to avoid climbs or descent and to correct the pitch of the nose when necessary. Another way to maintain straight-and-level flight is to visually check both wing tips of the aircraft and make sure they are equidistant from the ground. A pilot can use ailerons to make the needed adjustments. Once an aircraft is in straight-and-level flight, very little control pressure is needed from the pilot if an adequate trim system exists on the aircraft.

Turns

To initiate a turn, the pilot banks the airplane's wings in the direction of the desired turn by using the aileron controls. Simultaneously, the pilot appropriately utilizes the other primary controls. All primary controls are necessary when turning an aircraft.

Essentially, turns are divided into three classes: shallow, medium, and steep. Shallow turns have a bank angle of around 20 degrees or less. Aileron pressure must be utilized to maintain a shallow turn. In a medium turn, the bank angle is between 20 and 45 degrees. Often, with neutral aileron control pressure, the airplane will maintain a medium turn. Steep turns have a bank angle of over 45 degrees. Counter aileron control pressure must be applied to avoid overbanking.

In constant airspeed turns at constant altitude, it is necessary to increase the angle of attack by increasing back-pressure on the elevators. Also, additional power is necessary to counter the loss of speed due to increased drag.

Climbs

To initiate a climb, the pilot pitches the nose up by applying control pressure to the elevators. When the nose is pitched up, various forces are affected. Drag is increased, which requires more thrust. In fact, an aircraft's climb is limited by the amount of thrust available.

Descents

When an airplane changes its flight path from level to downward, a reduction in induced drag occurs. This requires a reduction in power to maintain airspeed.

Flight Envelope

In aerodynamics, the flight envelope (also known as service envelope or performance envelope) refers to an aircraft design's capabilities in terms of airspeed, altitude, maneuverability, and/or load factors. When an aircraft is pushed beyond its designed limitations, it is considered to be operating "outside the envelope." Operating outside the envelope is considered risky and dangerous. The expression "pushing the envelope" originated from military pilots taking an aircraft to its extreme limitations.

Commercial aircraft have smaller flight envelopes that are regulated by the Federal Aviation Administration (FAA) to ensure more safety for civilians during flight. Military aircraft have much larger flight envelopes, allowing more maneuverability and higher speeds. Extremely large envelopes are needed for some military aircraft since they serve a different purpose than simply transporting people from one city to another.

The following are a few of the factors that determine an aircraft's flight envelope:

Maximum airspeed is reached as air resistance gets lower and lower as a higher altitude is reached. At a certain altitude, oxygen levels are not high enough to feed the engines. This is where maximum airspeed is reached.

Stalling speed is the minimum airspeed at which an aircraft can maintain level flight. As the aircraft gains altitude, the stall speed increases.

Service ceiling is the maximum altitude of aircraft, determined by performance and wings.

Important Airport Information

Airports vary in shape, size, and quality. Pilots should educate themselves on the specific airports they will be operating within, but some common methods are universally accepted. Having common methods that are adopted by most airports simplifies communication and reduces the possibility of mishaps. Among other airport information, air traffic control and pilots should have a common understanding of lights, surface markings, and signs.

Taxiways are paved routes on which planes travel (taxi) to and from runways (where planes land and take off). Typical taxiway signs include the following:

Sign	Lettering Color	Sign Color	Meaning
Taxiway Location	Yellow	Black	Identifies taxiway on which aircraft is located
Runway Location	Yellow	Black	Identifies runway on which aircraft is located
Taxiway Direction	Black	Yellow	Defines direction and designation of taxiway intersections
Runway Exit	Black	Yellow	Identifies the exiting taxiway of the runway
No Entry	White	Red	Identifies areas where aircraft entry is prohibited
Runway Mandatory Instruction	White	Red	Denotes entrance to runway or critical areas
Runway Distance Remaining	White	Black	Indicates the runway distance remaining in thousands of feet

Common types of airport lights:

- *Runway edge lights:* White lights that outline the edges of a runway during darkness or periods of low visibility. These lights can be low, medium, or high intensity.
- *Runway end identifier lights (REIL):* Two synchronized, unidirectional flashing lights denoting the end of the runway.
- *Runway centerline lighting system:* Lights in the centerline spaced every 50 feet. They start out as white, then alternate between red and white, then are solid red toward the end of the runway.
- *Touchdown zone lights:* To indicate the touchdown zone in adverse landing conditions, two rows of transverse light bars are installed symmetrically about the runway centerline.
- *Taxiway centerline lead-on/lead-off lights:* Alternating green and yellow lights are sometimes installed to aid pilots in entering/exiting the runway onto/from the taxiway.
- *Land and hold short lights:* Pulsing white lights installed across the runway at the hold short point.
- *Taxiway edge lights:* Blue light or reflectors used to outline the edge of the taxiway during darkness or low visibility.
- *Taxiway centerline lights:* Green lights installed along the centerline of the taxiway.
- *Clearance bars:* Three yellow lights to increase visibility of a holding position and to indicate the location of an intersecting taxiway.
- *Runway guard lights:* Runway guard lights are installed to point out taxiway/runway intersections. They are a pair of yellow flashing lights on each side of the taxiway. They can be a row of in-pavement yellow lights across the taxiway at marked and signed holding points.
- *Stop bars:* A row of red lights across the taxiway across the holding position. Air traffic control will turn the stop bar lights off and the taxiway lead-on lights on with the issuance of clearance to proceed.
- *Approach lighting system:* A lighting system consisting of lightbars, sequenced flashing lights, or a combination of the two that lead up to the approach end of the runway.

Visual Approach Slope Indicators (VASI)

The visual approach slope indicator (VASI) typically consists of two light bars located on the left or right side of the approach end of the runway. When the farther bars are red and the closer bars are white, it indicates that the airplane is properly angled for approach. When both light bars are white, it indicates that the airplane is flying too high over the proper glide path. When both light bars are red, it indicates that the airplane is flying too low below the proper glide path. These lights are visible 3-5 miles away during the day and 20 miles away at night.

Three Basic Types of Runways:

Visual runways: Typically in small airports, visual runways have no lighting and few markings. The pilot must see the ground to land, as the use of instruments is not sufficient for landing.

Non-precision instrument runways: Often used at small to medium airports, non-precision instrument runways can provide horizontal positioning guidance as the airplane approaches. Depending on the surface, these runways can have threshold markings, designators, centerlines, and sometimes an aiming point (approximately 1000 ft from the landing threshold).

Precision instrument runways: Typically used at medium to large airports, precision instrument runways have an instrument approach procedure using a landing system that provides both horizontal and vertical guidance. These runways have thresholds, designators, centerlines, aiming points, blast pads, stopways, and touchdown zone marks.

Helicopters

Helicopters, or choppers, are a type of rotorcraft in which lift and thrust are supplied by rotors. The rotor wings, or rotor blades, are attached to a mast. Helicopters are able to accomplish missions that other aircraft cannot. Some of the capabilities that set a helicopter apart from a fixed-wing aircraft are the following: landing and taking off vertically, flying at slower speeds, hovering over fixed locations, and flying not only forward but also backward and laterally.

Most modern helicopters share the same major components: airframe, fuselage, landing gear, powerplant, transmission, main rotor system, and tail rotor system. The airframe is the fundamental structure of the aircraft and is attached to major components. The fuselage includes the cabin where the pilot, crew, and cargo are housed. The landing gear could be skids, wheels, skis, or floats. The powerplant is its engine. The transmission transfers power directly to the main and tail rotor systems. The main rotor system provides the aerodynamic forces necessary to fly. The tail rotor system is an anti-torque component that keeps the helicopter from turning due to torque.

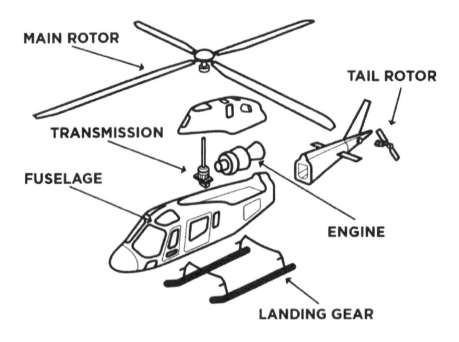

Helicopters come in many shapes and sizes:

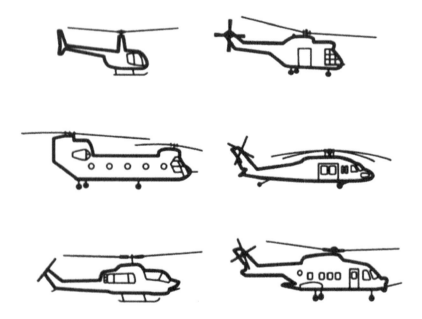

Helicopters encounter the same four forces as any other aircraft: lift, weight, thrust, and drag. In addition, helicopters are subject to other unique forces:

Torque Compensation

Newton's third law of motion states, "To every action, there is an equal and opposite reaction." As the main rotor rotates in one direction, the body of the helicopter has a tendency to move in the opposite direction. Typically, helicopters have a tail rotor to counteract this torque effect. The tail rotor will push against the aircraft's tail to compensate for torque.

Translating Tendency

When a helicopter is hovering, it has a tendency to drift laterally to the right. This is caused by the lateral thrust supplied by the tail rotor. The pilot may prevent lateral drift by tilting the main rotor to the left, which counteracts the tail rotor thrust to the right.

Centrifugal and Centripetal Force

The rotor system produces centrifugal force and centripetal force. Centrifugal force has a propensity to make rotating bodies move away from the center of rotation. Centripetal force counteracts this centrifugal force by keeping rotating bodies a certain radius from the center of rotation.

Coriolis Effect

The Coriolis effect is also known as *the law of conservation of angular momentum.* It states that an object will have the same velocity of momentum unless acted upon by an external force. Angular momentum accelerates as a rotating body moves closer to the axis of rotation. Imagine a spinning ice skater with her hands close to her body. Assuming her body spins at the same speed, her hands will move at a higher velocity as she fully extends her arms and hands. The diameter of the turn increases (extended arms), but her rotations per minute stay constant (her body is spinning at the same speed). Her hands are traveling a longer distance, resulting in an increased velocity to maintain the same revolutions per minute. As rotor blades move around their axis of rotation, there is a slight change in diameter due to the blades' flapping. The change in diameter creates a small acceleration and deceleration of the rotational velocity, resulting in fluctuations of lift.

Gyroscopic Precession

Gyroscopic precession is the phenomenon in which force applied to a rotating object is manifested 90 degrees later in the direction of the rotation.

Induced Flow (Downwash)

When rotor blades have a flat pitch, no induced flow occurs. Instead of being deflected downward, airflow moves behind the trailing edge of the blade at the same angles as it hits the leading edge. As blade pitch angle increases, the rotor system creates a downward vertical movement of air that is often referred to as downwash. Induced airflow increases as the rotor blade angle of attack increases. Induced airflow is greatest during a hover with no oncoming wind because there is no horizontal airflow affecting the rotor disk.

Effective Translational Lift

A hovering helicopter takes a lot of power. The efficiency of the rotor system is improved with an increase in horizontal airflow. As the helicopter moves forward, the airflow is pushed backward. Effective translational lift is a term used to describe the speed at which the rotor system realizes the benefit of the horizontal airflow. This happens as the helicopter moves out of its own downwash and into undisturbed air. Depending on the aircraft, this typically happens between 16 and 24 knots of airspeed.

Transverse Flow Effect

In forward flight, airflow passing through the rear portion of the rotor disk has a greater downwash angle than airflow moving through the front portion. The difference in airflow between the fore and aft parts of the rotor disk is known as transverse flow effect. This causes unequal drag in the fore and aft parts of the rotor disk, resulting in vibrations that the pilot can easily recognize.

Dissymmetry of Lift

As the helicopter moves forward, the relative airflow on the advancing halves of the rotor blades is different from the relative airflow on the retreating halves of the rotor blades. This creates a difference in lift between the advancing and retreating rotor blades. This difference is known as dissymmetry of lift. The pilot uses blade flapping and cyclic feathering to compensate.

Helicopter Controls

Three major controls are used to handle a helicopter: collective pitch controls, cyclic pitch controls, and anti-torque pedals (tail rotor control). In addition to these major controls, throttle control is used to fly the helicopter.

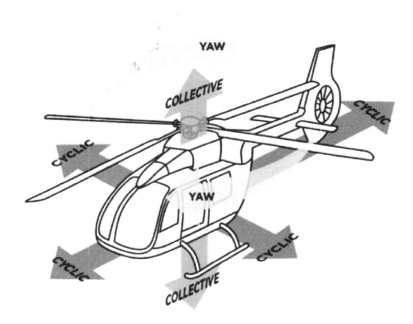

Collective Pitch Control
Collective pitch control is used to change the pitch angle of the main rotor blades simultaneously ("collectively," as its name implies). As the pitch angle of the blades increases, more drag occurs, which slows down the speed of the blade or revolutions per minute (rpm). By decreasing the pitch of the main rotor blades, less drag occurs, which speeds up the blade rpm. Faster blade speed helps generate lift. Hovering occurs when lift and weight are equal.

Cyclic Pitch Control
Cyclic pitch control can fly the helicopter forward, backyard, left, or right. If the cyclic control (sometimes just called *the cyclic*) moves forward, then the rotor disk (the circular plane through which the rotor blades pass) tilts forward. If the cyclic control is moved to the right, the rotor disk tilts right because the rotor disk tilts in the direction of the cyclic control. The cyclic control can control each blade independently. As each blade passes through the same point in the cycle, each blade has the same pitch angle.

Anti-Torque Pedals (Tail Rotor Control)
Anti-torque pedals serve a purpose similar to rudder pedals in a fixed-wing aircraft. The pedals are often located in the same position. They are used to change the pitch of the tail rotor blades, increasing or decreasing tail rotor thrust in a desired direction. The nose of the aircraft will yaw in the direction of the applied pedal.

Throttle Control

The throttle controls the power of the engine and ensures that the rotational speed of the rotor blades is in a desired range. The throttle is often a part of the collective control lever. It is usually controlled by twisting left for increased rpm and twisting right for decreased rpm.

Before You Start the Practice Questions

On the next page, you will begin an Aviation Information practice test. Set a timer for 8 minutes before you start this practice test. Giving yourself 8 minutes will give you an authentic feel for how long you have to finish this portion of the AFOQT. Like the official test, this practice test has exactly 20 items. Set your timer and begin.

Aviation Information Practice Test Questions

(8 Minutes)

1. In a hover, which control moves the helicopter forward, backward, or sideways?
    a. Collective
    b. Coriolis
    c. Cyclic
    d. Anti-torque pedals
    e. None of the above

2. Which of the following is considered a primary flight control?
    a. Rudder
    b. Fuselage
    c. Trim
    d. Lift
    e. Wing

3. Which of the following is NOT one of the four main forces that act on an aircraft?
    a. Lift
    b. Gravity
    c. Drag
    d. Wind
    e. Thrust

4. A component of the propulsion system that generates the energy necessary to propel an aircraft is known as the
    a. Fuselage
    b. Powerplant
    c. Torque
    d. Piston
    e. Throttle

5. Which is NOT a part of the tail assembly?
    a. Elevators
    b. Rudder
    c. Trim tabs
    d. Ailerons
    e. Vertical stabilizers

6. Which refers to the orientation of an aircraft with respect to the horizon?
    a. Altitude
    b. Angle of attack
    c. Attitude
    d. Descent
    e. None of the above

7. Which of the following is NOT a primary flight control?
    a. Elevator
    b. Rudder
    c. Aileron
    d. Spoiler
    e. None of the above

8. Which of the following BEST describes an airfoil?
    a. The movement of the aircraft in relation to the angle of attack
    b. The main structure of an aircraft where the cockpit, cabin, and cargo exist
    c. Resistance that is inevitable during flight
    d. A primary control of the aircraft to help counteract yaw
    e. The surface that is designed to help generate lift and control of an aircraft by handling airflow

9. When is an airplane considered to have tricycle landing gear?
    a. When the third wheel is near the tail
    b. When the third wheel is near the nose
    c. When all three wheels are near the tail
    d. Options A and B
    e. None of the above

10. The basic weight of an aircraft includes
    a. Crew
    b. Pilot
    c. Cargo
    d. Flight controls
    e. Bombs

11. Along which axis does roll occur during flight?
    a. Longitudinal
    b. Lateral
    c. Vertical
    d. Options A and B
    e. All of the above

12. Which of the following maneuvers requires increased throttle and back-pressure on the elevator?
    a. Straight-and-level flight
    b. Turning
    c. Climbing
    d. Descending
    e. None of the above

13. A stall occurs when an aircraft experiences
    a. Increased air resistance and decreased lift
    b. Decrease in power and altitude
    c. Crosswind while banking
    d. Options A and C
    e. Too little air resistance

14. Ground effect is
    a. Increased drag and decreased lift from flying close to the ground
    b. Electricity flowing through the fuselage
    c. Gravity increasing the weight while landing
    d. Reduced drag and increased lift from flying close to the ground
    e. The impact on an aircraft while landing

15. The chord line is the imaginary line between
    a. The leading and trailing edge of the airfoil
    b. The rudder and the elevator
    c. The trim and the vertical axis
    d. The horizontal stabilizers and the vertical stabilizers
    e. The direction of flight and the center of gravity

16. The Coriolis effect is when an object moving in a rotating system experiences a force that is
    a. Perpendicular to the direction of motion and the axis of rotation
    b. Parallel to the direction of motion and the axis of rotation
    c. Close to the direction of motion and the axis of rotation
    d. Vertical to the direction of motion and the axis of rotation
    e. Diagonal to the direction of motion and the axis of rotation

17. Which option BEST describes effective translational lift?
    a. Difference in lift between retreating and advancing blades of the helicopter
    b. Difference in airflow between the aft and forward parts of the rotor disk
    c. Vibrations that increase with altitude
    d. Improved efficiency that results from directional flight
    e. Increase of pitch at all points of the rotor blade simultaneously

18. How far away can an approaching plane see the visual approach slope indicators during the day?
    a. Up to 20 miles
    b. Up to 8 miles
    c. Up to 10 miles
    d. Up to 2 miles
    e. Up to 5 miles

19. What option BEST describes a taxiway?
    a. The area in which passengers are transported to and from the airplane
    b. The paved route on which an aircraft travels to and from the runway
    c. The visual cue from the air traffic controller to signal that an aircraft can proceed
    d. The area in which an airplane lands and takes off
    e. None of the above

20. Which of the following is a fundamental flight maneuver?
    a. Yaw
    b. Pitch
    c. Bank
    d. Climb
    e. None of the above

Answer Guide to Aviation Information Practice Test

1. **C.** The cyclic control is the stick that controls a helicopter's movement forward, backward, or sideways. It also has various buttons that control radio, intercom, and trim.
2. **A.** The rudder is a primary flight control.
3. **D.** The four main forces that act on an aircraft are lift, gravity, drag, and thrust.
4. **B.** The powerplant of an aircraft is its engine (or engines). This is a component of the propulsion system that generates thrust to propel the aircraft.
5. **D.** The tail assembly of an aircraft usually includes the tail cone, horizontal and vertical stabilizers, elevators, and rudder. Ailerons are located on the aircraft's wings.
6. **C.** Attitude refers to the aircraft's orientation with respect to the horizon.
7. **D.** The spoiler is a secondary flight control. Elevators, rudder, and ailerons are primary flight controls.
8. **E.** The airfoil of an aircraft is a surface designed to help generate lift and control the aircraft by handling airflow.
9. **B.** If an airplane has tricycle landing gear, the third wheel is near the nose.
10. **D.** The basic weight of an aircraft is the weight "as built." The weight of the structure, installations, furnishings, systems, and powerplant are considered part of the basic weight. Crew, pilot, cargo, and bombs would be things added later and not considered part of the basic weight of the aircraft.
11. **A.** The movement of roll or bank occurs on the longitudinal axis of the aircraft during flight. Yaw occurs on the vertical axis and pitching occurs on the lateral axis.
12. **C.** Climbing requires the simultaneous increase of throttle and back-pressure on the elevator.
13. **A.** A stall occurs when the aircraft's wings experience increased air resistance and decreased lift.
14. **D.** When a fixed-wing aircraft flies too close to the ground (or another fixed surface), the result is an increase in lift and decrease in drag.
15. **A.** The chord line is an imaginary line that runs through the leading and trailing edges of the airfoil.
16. **A.** The Coriolis effect occurs when an object moving in a rotating system experiences a force perpendicular to the direction of motion and the axis of rotation.
17. **D.** Effective translational lift is the improved efficiency that results from directional flight.
18. **E.** An approaching plane can see visual approach slope indicators (VASI) up to 5 miles away during the day. During the night an approaching plane can see VASI from up to 20 miles away.

19. **B.** Taxiways are paved routes on which planes travel (taxi) to and from runways.
20. **D.** Climbing is a fundamental flight maneuver. The four fundamental flight maneuvers are straight-and-level, turn, climb, and descent. Yaw, pitch, and bank are movements controlled by primary flight controls (elevators, ailerons, and rudder).

EXAM SUCCESS 2.0

10 Tips for Successful Test Preparation

No matter what the stakes, goals, or consequences involved, major standardized and professional tests can cause a lot of anxiety, heartache, and frustration. While some people are natural test-takers, many find some combination of the time constraints, test format, and subject matter to be overwhelming and discouraging.

Part of the reason that major tests seem so intimidating is that standardized tests are designed to elicit stress. Fortunately, there are tried and true techniques for preparing for testing that can help maximize your scores and minimize your anxiety. While some of these test-taking strategies are seemingly common sense, others are often unknown or forgotten by test-takers.

Like any other high-stakes performance or situation, the trick to successful test-taking is a preparation regimen and strategy that helps ensure you are as ready as you can be to give the test your full effort, concentration, and focus. With this in mind, we compiled our top ten most effective tips for test-taking preparation. While these can't guarantee a high score, by following this guide, you can put yourself in the best possible position to succeed.

Tip #1: Learn More About the Test Format – Know Where the Trouble Spots Will Be Before You Take It

Every major test, from the ASVAB to the AFOQT, is designed not only to assess content knowledge, but also gauge how you function in high-pressure situations designed to push your abilities. Each individual section of a major test is formatted to force test-takers to show not only how their brains perform, but how they handle budgeting their time and effort.

This brings us to our first tip, which is beyond simply knowing what each section of the test consists of; it's about knowing how many questions each section includes, what types of questions or categories of questions you can expect, and how much time you have on average to complete an individual question. While this may seem silly, understanding time constraints and question types will help you determine which parts of the test will be most difficult for you, so you can streamline your test preparation time, choose the practice questions you complete beforehand, and seek out any extra help you may need.

There are two effective ways to glean this information: first, the test creator always has information about the test available online, so search this out; second, take time to look over and complete a practice test at your own pace. The more you know what to expect, the better you can structure your preparation.

Tip #2: Take a Practice Run Well in Advance

As soon as you know that you have, say, the AFOQT on the docket in a month or two, take a practice test! There are numerous resources available, both online and in test prep books available through Amazon or any book seller, and plenty of practice tests for you to choose from. Find one and put yourself through a full simulation of the test, complete with official time constraints, as soon as possible.

There are two major benefits to this. The first is what we outlined above: the diagnostic value of understanding what content was difficult, which categories of questions flustered you or require more preparation, and whether you are pressed for time in specific sections.

The second major benefit of practicing well in advance is that it lets you create an effective prep calendar. Whether you have some extra funds for test preparation courses or a tutor, or will be handling all test prep yourself, an effective preparation calendar is key to helping you maximize your results and experience growth in your areas of weakness before the big day arrives.

Tip #3: Create a Test Preparation Calendar and Schedule, and Stick to It

This tip isn't as simple as marking a few days in your schedule to take practice exams. Much like preparing for a big game, you want to think of test prep as focusing on areas of weakness that can be improved. Unfortunately, many people make a major mistake with test preparation, especially if they received a low score on their first practice run; they try to improve every area of weakness simultaneously. This is counterproductive, as it is not realistic (especially if you only have weeks to prepare), and it can often leave you feeling more overwhelmed or discouraged than before.

Instead, consider either A) the areas of difficulty you will need to cover between now and the test, or B) the amount of prep material (say, chapters in a test prep book) that you will need to get through. For option B, if you have 15 chapters of test prep content to cover in

5 weeks, setting deadlines of 3 chapters each week can ensure you actually cover the content at a realistic and reasonable pace. For option A, if you have the same 5 weeks and 10 categories of questions throughout the test to improve on or review further, spend each week focusing on 2 categories of questions to ensure your prep is comprehensive.

Remember, if you are pressed for time, it may not be possible to cover every single thing you need more help or practice with. Don't worry! The key is to be realistic about your prep calendar, which means that the areas of biggest need come first. Use your practice scores and experience to determine this. If there are a few categories in which you missed 1-2 questions, but one category in which you got everything wrong, you will have much more success focusing on the area of greater need instead of spreading yourself too thin.

Tip #4: Identify Test Preparation Resources That Are Right for You

First, consider what your biggest struggles with the practice test were. Was it the content or certain areas of content? Was it running out of time? Was it dealing with general anxiety or stress that made it hard for you to concentrate and think clearly? Each of these are distinct issues (even if you experienced all three), and as such, the resources required for improving in each of these areas are also distinct.

While test-taking classes and private tutors are certainly helpful for any of these problems, the costs associated often make them prohibitive for many students. Instead, here are some suggestions for each of these three main problem areas and what resources you should consider to help with each:

- **Time Allocation** - Quite simply, what you need are more practice tests or practice test sections to help improve your sense of when you need to move from question to question. We also recommend doing practice tests or questions with slightly *less* time than the actual test will give you. While this might lead to some struggles with your early practice, by training yourself to work with less time than the test provides, you will feel less pressured on the actual test day.
- **Content Issues** - This is the broadest area of need, but also offers the widest array of test prep resources, from textbooks and online guides, to simply seeking out help from content-area teachers (say, your English or math teacher if you are practicing to take the AFOQT while in school). Follow the same approach as above: *do not try to fix everything*. Focus on the areas of biggest need, and concentrate on resources in those areas.
- **Test Anxiety and Stress** - Let's talk more about that with Tip #5...

Tip #5: Put Yourself in the Best Physical and Mental Condition to Combat Test Anxiety

We know that the idea of going to the gym to prepare for taking a test seems a bit silly or counterintuitive, but it's actually backed by science! People in better physical condition have been proven to think more clearly, maintain focus and retain content information more effectively, and are able to remain calm and relaxed more easily. While test anxiety can affect everyone differently, and while everyone should expect some normal stress and anxiety before any big test, some people find these issues to be crippling to their concentration and subsequent results. Here are a few general tips to help put yourself in the best position to succeed:

- **Surround yourself with positive people and influences** - If you prepare with mentors or peers who are consistently negative (whether about your chances or their own), you will find it hard to ignore the negative voices in your own head. If you are practicing test strategies with peers, focus on the gains you are making, not the struggles. If you have friends or family who are voices of doubt around you, ask them to refrain from making negative comments, or find a quiet and positive space away from them to do your preparation work. It's a simple, effective tip that's proven by psychology!
- **Get more sleep well before test day** - High school and college students are especially susceptible to sleep deprivation and exhaustion that can affect even the most effective brains. While everyone knows you should sleep well the night before a test, beginning that habit weeks in advance is much more effective. Since your brain takes time to acclimate to (or recover from) a new sleep schedule, give your body time to adjust before the big day arrives.
- **Eat well and work out** - If you are already doing this, keep doing so; you probably know that you have more energy and concentration from taking care of your body. Even if you aren't physically active, simply walking instead of driving and cutting out fried foods and sugars for a few weeks before the test can give your brain the go-power it needs to bump your score up a few points.
- **Practice, practice, practice** - While you may be frightened or stressed out by the test itself, the more experience you get with taking it, the less surprised you will be by anything on test day. Anxiety is often driven by the fear of the unknown, so the more you can expose yourself to the test in advance, the more you will know what's coming.

Tip #6: Don't Fall Victim to False Beliefs About Testing

There are many cliches and rumors about major tests that are easy to fall victim to: there's no way to improve your score that much after the first time, some people are "just bad test-takers," etc. False beliefs are spread because they offer people easy excuses for struggles or frustrations with a test, but they are false for a reason. Few things are universally true about succeeding or failing on a major test, so the more you can focus on your own experience, the more you can actually improve, instead of letting others' beliefs guide your preparation the wrong way.

Tip #7: Use Rewards to Motivate Your Preparation Work

There's a reason standardized tests are so stressful and challenging—they are hard! As such, the prep work you need to do for them is hard as well. The test preparation period can be not only tedious, but also discouraging if you find your results seeming to plateau in a specific content area, but the only way to improve is to keep at it. So how can you help yourself do this?

Quite simply, you should consider a point from Psychology 101: every hard job is easier to do if there is an incentive for completing the work. Consider what your own personal rewards might be. Is it getting dinner at your favorite restaurants with your friends? Spending an hour on Netflix enjoying your favorite show and giving your brain a rest? Whatever those motivators may be, work them into your test prep calendar.

This is also important for keeping your mind healthy and fresh. A common mistake by test-takers is to overwhelm themselves by attempting to prep for everything, while forgetting to give themselves time to refresh their minds. By structuring these little rewards throughout your test prep calendar, you not only help motivate yourself through those different deadlines for test prep, but also ensure your brain is in top shape when test day arrives.

Tip #8: Maximize Your Retention Abilities

While some of the previous tips are key to this step as well, the way that you attempt to remember and organize information in your brain is also key to ensuring you get the most out of all the test prep work you do. Here are a few tips for retaining important information more effectively:

- **Use associations and practical applications** - Even the most mundane math concepts often have practical applications and associations that make things like formulas easy to remember. For example, the phrase "Please Excuse My Dear Aunt Sally" is used to remember the order of operations in math. Always look for examples or uses in your own experience to connect test concepts you are focusing on. This is psychologically proven to make recall easier, even in a high-pressure situation.

- **Try to use vocabulary that you are learning** - Often, an extensive vocabulary is one of the key factors for test success in any content area, so test-takers spend large portions of their prep memorizing common vocab terms or phrases. However, memorization alone isn't usually enough. To truly make your vocabulary work useful, employ the words you are using in letters, homework assignments, or even a personal journal. By actually using the terms you are learning, your recall will be much easier.
- **Narrow your focus areas** - If you try to cover everything, you'll often have trouble remembering anything. So perform triage as detailed above and choose a few key areas to focus on. Even if you still have some smaller issues with certain categories of questions or content areas, you'll at least retain information in your areas of greatest need.

Tip #9: Do Not Let Test-Prep Sabotage Other Areas of Your Life

This tip sounds like common sense, but people who are feeling overwhelmed by a test will often allow themselves to be consumed by test preparation. This can happen at the cost of normal school work, work life, or family life. Unfortunately, that can lead to a cycle that makes everything worse; as anxiety grows from unfulfilled obligations outside of test prep, your anxiety about test prep itself gets worse.

So be logical with your calendar, and remember that test prep cannot come at the expense of your other responsibilities, both academic and social. While you can't control the exact questions or content matter on the test, you can control your environment and personal life to put yourself in the best position to succeed.

Tip #10: Remember That It's Just a Test

We're not kidding. While every major standardized test is important and high-stakes for a reason, at the end of the day, it's still simply a test. It's not life or death, or the only thing that gives your life value. And remember, even if you have a bad experience the first time, you can use that bad experience to improve your prep calendar for your second attempt, which is often when people find the most success on any major test, be it the SAT or AFOQT. So don't put more pressure on yourself than you need to. You will be fine!

10 Tips for Successful Test-Taking

No matter how much or how little you have prepared for it, the big test day will arrive before you know it. While you can do plenty ahead of time to put yourself in the best position to succeed, once the booklets are handed out and the test timer begins, the pressure ramps up and your approach to the test will help determine your success. While content knowledge and preparation are a major part of the equation, there are a number of methods and tips to follow when taking any major test in order to put yourself in the best possible position to earn a high score. Here are ten of the most important tips to follow to give yourself the best chance to maximize your score and get the most out of your testing experience.

Tip #1: Time Management Is Key

Most standardized tests are specifically designed to put you in a high-pressure situation, not only in the total number of questions you are expected to answer during a specific block of time, but also in the difficulty of some of the questions. If you spend too much time trying to crack a difficult question, you may find yourself running out of time before even getting to see the last few questions in a particular section. This is especially frustrating if you miss out on easy opportunities for points just because you got stuck spending too much time on a particular problem.

The solution to this is easy: while you always want to answer every question, you must force yourself to make a guess and move on if you find yourself spending more than a minute or two on a particular question. With most standardized test sections giving you a maximum of 60-90 seconds on average per question, just a few delays can cost you more points than simply skipping one hard question. Moving on from particularly difficult questions also ensures that you don't get too frustrated or discouraged by a single section, especially when major tests usually involve three or more banks of questions to work through. If you have time at the end of the section, you can go back to these questions and give them more thought.

Tip #2: Eliminate the Wrong Answers First

Another efficient method to save time and increase your chances of getting the correct answer, even if you are unsure, is to use the process of elimination. Aside from being a general good practice, there is a purely statistical reason for using this strategy to aid your test taking. If your question has four answer choices, and you can rule out two obviously wrong answers before the two that you are trying to decide between, you've upped your odds of getting the right answer from 25% to 50%, content knowledge aside. Even if you are choosing from five possible answers, eliminating two takes you from a

20% to a 33.3% chance in choosing the right one. Eliminate three, and you've gone from 20% to 50%, more than doubling your chances of your best guess being correct. Playing the odds like this significantly increases your ability to make an effective guess.

Tip #3: Read the Entire Question and Answers; Don't Skim

Whether in a high-pressure situation like a major test, or even in an environment where you have ample time to make your choices, people are psychologically predisposed to see a few words they know and skip directly to answering a question by assuming they know what the question is looking for. Test-makers tend to try to exploit this; a question may start out seeming to ask for one thing, but clarifies what it is in fact asking for later on. This means that if you are skimming, you may miss a crucial part of the question. One of the answer choices is usually designed specifically to penalize people who do this.

The longer the answers are, the more likely there are some tricks or twists hidden in the choices. Instead of simply trying to diagnose the differences between the answers quickly, carefully read each answer choice to ensure you aren't missing anything critical before you make your final decision.

Tip #4: Trust Your First Instinct

It's perhaps one of the biggest cliches in test-taking advice, but there is ample psychological research that has proven that the brain's first instinct at the right answer is often the right one. This is true on a subconscious level, in that your brain may have synthesized information that you aren't even consciously recalling, but that has been buried deep in your memory. With this mind, always go with your gut if you are having trouble choosing, as research proves that this is probably the right choice anyway. Note that if you have a clear and logical reason for changing your mind on an answer (maybe you read the question wrong the first time or later recalled an important fact), it is okay to go with your second choice; however, do not change an answer simply because you have vague doubts.

Tip #5: Avoid Overthinking the Question

Long and complicated questions (often paired with long and complicated answer choices) often flummox test-takers, especially those worried about finishing a section under the time limit. These tests are designed to cause and exploit this particular type of anxiety, so don't let that happen to you. When faced with a question that seems vague or ambiguous, focus on what the question is asking you to do, not your perceived intent or potential alternate interpretations of the question.

You can use the answers as a means of clarifying this for yourself; if the answers are all similar (many questions will have several similar answers, specifically to exploit a particular small difference), that should give you a clue as to what the question is asking of you. While you can't avoid difficult or vague questions on any major test, you can simply keep calm, take some time to sort out what the question is really asking you, and maximize your chances at the correct answer by keeping your analysis of the question simple.

Tip #6: Keep an Eye Out for Subtle Negatives

One of the oldest tricks in the book for test designers looking to confuse or throw off test-takers is to include what is known as a "subtle negative": words like "not" or "except" that can drastically change the thrust of a question and mislead the test-taker if they miss them. You can see these tricks in questions that are structured similarly to "Which of the following is not..." which warps the common question stem "Which of the following is..." These question types are common on all forms of standardized tests, no matter the goal or particular subject matter. Despite the fact that this is a widely known ploy, it still catches many unsuspecting or rushed test-takers off guard and costs them precious points. If you catch one of these subtle negatives, be sure you are clear on what the question is asking before you choose the answer.

Tip #7: Take Note of Qualifying Words

Similar to subtle negatives, another test-maker trick is to include what are known as "qualifiers" or "hedges." These questions or answers commonly include words like "sometimes," "often," "almost," "most," and other similar words that indicate there are some exceptions to the concept being tested. These should immediately clue you in to two things: first, that there are some elements that fall outside of the rule being questioned, which means there may be a trap at work. Secondly, hedges or qualifiers are often used in conjunction with answers that assess whether a test-taker has spotted them; in other words, one of the answer choices will be there specifically to test whether you noticed the qualifier. This is in contrast to definitive words like "every," "always," and "all," which make it easier to eliminate answers based on whether they fit the rule or not. While tests often include questions of both forms, if you aren't careful in noticing the differences, you can fall victim to this trick fairly easily.

Tip #8: Reword the Question into Simpler Terms

Many questions cause difficulty for test-takers not necessarily because of the subject matter, but because of overly complex language used to ask the question. Aside from consuming your valuable time and potentially confusing you even if the concept being tested is simple, these questions are also meant to put you under even more pressure.

The easy fix to this situation is to reword the question in plain language, ensuring you are clear on what you are really being asked to figure out. While you may not know every word or phrase being used, in order to be fair, the question will include key words to guide you to the correct answer. By rewording the question, you can make it easier to begin your process of eliminating the wrong answers and finding the best one.

Tip #9: Don't Fall Victim to Patterns That Aren't There

One of the most common mistakes you can make is reading into your sequence of answers and assuming that because you have chosen a particular answer three times in a row, or spotted some other pattern (such as A-B-B-A) on your answer sheet, you have erred in your choices. While it is certainly rare to see four or more answers in a row that are correct, it is not unheard of, and sometimes the random ordering of questions when a test is being created can simply lead this to happen by chance. Never let your answers be determined by the choice that came before or after. This is very similar to the Gambler's Fallacy, where people inaccurately attribute odds to the next roll of a dice even if the odds of each roll never changes. Trust that your answers are correct based on your knowledge of the content, or by following the appropriate strategies for eliminating options and choosing the best remaining answer. Relying only on patterns can lead to incorrect choices, and will often consume valuable time and mental energy that you could instead be spending moving on to remaining questions.

Tip #10: Keep Calm and Carry On

It's an overused saying, but an especially true one for high-stakes testing environments. The easiest way to forget all of the preparation and testing strategies you've been working on is to allow yourself to become panicked by a few tough questions in a row, or discouraged because you haven't been sure on a number of questions. Panic and stress can only further decrease your chances at success, so finding strategies to avoid them is critical before you begin your work. While time is certainly an issue, taking a moment to breathe, close your eyes, and forget about the test can be a simple way to get yourself re-centered and refocused before you continue. Don't underestimate the importance of this. Tests are designed to cause and exploit anxiety. Don't let them get the best of you!